Otto von Bismarck, the great German chancellor a century ago is reputed to have said that making sausage and politics was best done out of sight. Government meat inspection is both politics and sausage.

A book not for the squeamish! A look under the big flat rock!

But a book aimed at provoking some laughter, some tears, and a great deal of thought.

Every government inspected slaughter plant is required to have veterinary inspection. Dr. Massie spent some twenty odd years working as a meat inspector prior to retiring a few years ago. He worked as a government veterinarian eradicating livestock diseases before that.

These are a series of short stories spanning a century, stories about meat inspectors, about what they do, what they think, and what happens to them. Some of them are true, some are fiction.

Read about where meat comes from, how animals are made into edible food, how money is made (and lost), and how government tries to keep everyone happy and healthy (if not totally informed.)

Reviewers have said:

"Fine tuned and well written."
Joe Vitale, author of *Turbocharge Your Writing*

"Very enjoyable."
Dr. Philip Dawson, Ivor State Laboratory

"I enjoyed reading the book very much. Writing is excellent."
Dr. Paul J. Friedman, former Director of Meat Inspection, Virginia Dept. of Agriculture

"Well Written"
George Nasea, retired kill floor foreman

$12.95
ISBN: 0-9638718-0-3

ISBN: 0-9638718-0-3
9 780963 871800
W9-DGR-489

To order more copies of

Jungle Tales

make a copy of the form below and
send it to us,
or call
1-800-968-7065

WILDWORKS PUBLISHING
8331 Silver Shadows Lane
Spring, Texas 77379

Ship to: _____

Name: _____

Address: _____

City, State, Zip: _____

Phone: _____

Fax: _____

☐ Check this box if you have special
instructions written on back of this form.

JUNGLE TALES is just $12.95 plus $5.00 shipping & handling per
book.

I would like_____ copy(s) of *JUNGLE TALES.*

Total amount including shipping _____

Orders will be shipped UPS to a street address only (Not a P.O. Box).
Texas residents, please add Texas sales tax of 8 ¼%.

Winfield Massie

Jungle Tales

Adventures in Meat Inspection

Written and Illustrated by
Dr. Winfield Massie

with
Apologies to
Upton Sinclair
and
Rudyard Kipling

Wildworks Publishing

Copyright © January 1995
Wildworks Publishing
8331 Silver Shadows Lane
Spring, Texas 77379

Library of Congress Number TXu 580 430
ISBN: 0-9638718-0-3

Printed in cooperation with Brockton Publishing
Houston, Texas
1-800-968-7065

This is a collection of short stories; the fictional adventures of meat inspectors, their coworkers and associates. Some stories are true. However, names and places have been changed to protect the guilty as well as the innocent. I hope that while the stories are fiction, larger truths will become evident to the reader as the stories unfold.

Rudyard Kipling first used the title and Upton Sinclair first explored this particular Jungle.

I wish to thank all those colleagues who stirred my imagination to write this narrative, including; P. F. Friedman, Carl T. Ellers, Mark Weisser, Joe Vitale, August Raber, P.J. Dawson, Robert Hardy, Davey Haverty, Jim Winterhalter, George Nasea, George Popovici, David Forbes, and John Foriest.

In memory of; W. W. Pittman, Herman York, Richard Mansfield, Moses Simmons, Warren Babcock, and C. R. Fitzgerald, O.R Rawls, Tom Monahan, and Elmer Fisher.

And thanks also to the cows and pigs who gave so much to further my career.

Winfield Massie, D.V.M.

Table of Contents

Confessions of a Meat Inspector

If you go to the hog pens in the wee witching hours between midnight and dawn, as I have, you will hear a low rumble like distant surf, or the murmur of an audience awaiting the opening curtain. The sound rises and falls and is sometimes punctuated by a crescendo of outraged squeals; "You took my seat!", or "That was my grain of corn!".

Unaware of impending doom, their concerns are immediate and trivial; or so it seems. Those in deeper thought say little.

The hours before dawn are conducive to strange dark thoughts involving life, death, purpose, destiny and all the rest that has bedeviled mankind since Adam ate of the Fruit of the Tree of the Knowledge of Good and Evil. It should be noted that growing up in litters began about the time that the Snake lost his legs and went underground to eat dust. So that both Man and Beast are afflicted with notions of social justice. One cannot discuss it with his fellow beast, but the hog for one will certainly protest injustice as he sees it.

Perhaps especially the hog, who is intelligent, born into a large gregarious family, and is by nature an opportunist and optimist. He is endowed with the same traits as the entrepreneur. He adapts to luxury, if it is available, but makes the best of the worst if forced to. To himself he says, "Tomorrow will be better." It is no coincidence that the businessman is symbolized by the pig. Both muddle through if they can, are outgoing, hate unpleasantness, are basically inoffensive, have a strong sense of property rights, and last but not least, are maligned by those who prey upon them. No doubt Islam and Judaism understood this when they removed the hog from their menu.

Traditionally, animals, especially hogs, were slaughtered in the very early hours of the morning when it was cool. Also the meat would be ready for sale at the market by mid morning the same day. Refrigeration has not really changed this. Starting in the wee hours has persisted. Perhaps bloody deeds are best done in darkness.

The pigs going to market are quite young. They have scarcely reached puberty. Hog castration persists but now is an anachronism, from a time when hogs were fattened for a year or more. Now they reach market weight in less than five months. Perhaps their youth explains their contentiousness. Old sows and boars seem much more civil towards each other.

Before slaughter, the animals are paraded forward and back so the inspector will see both sides. The animals are observed at rest as well. Some few will be dead on arrival. Some will die in the pens, sometimes from disease, nowadays more often from injury and stress. These will be accounted for and their dead carcasses will be sent to the rendering tank house where they will be made into grease and tankage. Others still alive but sick or crippled are marked for more detailed examination later... or they may be killed immediately and then join those already in the tank house. The government veterinarian must make these final judgments. One thinks of a small prayer for Dr. Mengele... and one's self. Certain similarities are probably more than coincidence.

It is not good to dwell too long on such macabre matters. The notion of being killed and eaten may seem horrid, yet we share this fate with the pig and all other living things. The elements that compose us are only lent to us for our lifetimes. It is a process that continues throughout our lives and ends after our deaths. Embalming means only a brief delay on the geological time scale. The water that washes through us during our lives has been in oceans, fish, clouds, dinosaurs, insects, plants, and will go on into others, animate and inanimate. And so it is with our other elements.

A living organism resembles a well regulated gas flame. It is a process disguised as an object. A complicated order is imposed upon matter, whether flame, man, or hog, and the elements that were once part of us pass through us and are then part of something or someone else. The elements remain, however dilute. It is we who come and go.

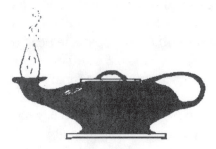

Primitive man, and woman, believed that what one ate had a bearing upon what one became. We still do, as both diet fads and nutritional science attest. Men sometimes believed that if you wanted to be wise then the way to wisdom was to eat a wise man or animal; or perhaps a strong one if that was your intent. Cannibalism was less often a matter of avoiding starvation than a ritual of respect. Nevertheless, since our flesh closely resembles that of the pig, prohibitions against the consumption of pork may be rooted in the avoidance of eating human flesh. It has been claimed that the peoples whom the Israelites fought in Canaan performed human sacrifice and ritual religious cannibalism like the Aztecs and Mayas.

Primitive hunters sometimes drove great predators off their kills. Just as a lion lurked at his kill and defended his property; so too would the invisible spirits that also killed man and beast alike defend their kills. Early man viewed an animal dying from a disease as having been killed by an invisible predator, who would also try to defend its kill until it had consumed what it wanted.

His idea was essentially correct, even though he did not know that the reason he could not see the predators was their small size and his lack of a microscope. Nor could he know that the predator's great power was in their vast numbers. He learned that fire would drive off the lion. It would also drive off the invisible predators. Perhaps not always but most of the time it would. We call it cooking.

Still, it called for the skills of a shaman, or priest, to exorcise, appease, or kill the evil spirits in dead animals, and to pronounce them fit for food. We still do. Islam and Judaism still involve their priests in animal slaughter, as do some other religions. We say grace over our food. It is the practice of hunters in Northern Europe to have a moment of silent prayer, or reflection when an animal is killed. It calls to mind the brevity of life, its beauty and its mystery, and that we can take life but not restore it.

Being descended from a long line of Scotsmen and Englishmen, I sometimes wonder if some of my far distant ancestors were Druid priests, and what they would think of my auguries. I too inspect the viscera of animals, and thanks to the ink we use, I too am often painted blue after a long day on the kill floor. I cannot often predict the future from a set of hog entrails but I can often see into that particular animal's past. I can usually see what diseases and parasites he has or has had, which in turn may tell what part of the country he came from. Sometimes I can see rural ignorance and poverty, sometimes cruelty and brutality...

I have finally reached market weight, 100 kilograms, I prefer to say. It has taken five decades rather than five months. In whimsical moments I wonder about my own dressing percentage, rate of gain, and conversion ratio. Perhaps if one should die as one has lived, I should have a simple tankside rendering service with quotations from the Manual of Procedures and from the Regulations..."the body which is corruptible being committed to the rendering tank, but the soul having been Inspected and Passed, shall be sent on to the Great Chill Cooler in the Sky..."

In any case it seems only right and natural to have sympathy for a fellow creature who is also fat and indolent.

In the interest of not writing too long a book, your respectful author has omitted some of the four letter words from the dialogue or abbreviated them. In the first place, some are used so often that they really aren't heard. In the second, there are proper terms for many of them which we should encourage people to use. There is the matter of using an obscenity for lack of a proper medical vocabulary. Consider the following conversation between a company trimmer and an inspector on the kill floor about a carcass which is contaminated by feces and must be partly skinned. After all the hog is dirty on the outside as well as the inside.

Trimmer: "What 'cha got here, Doc?"

Inspector: "Fecal contamination on the fore feet!"

Trimmer: "What! I already cleaned up that one."

Inspector: "Well you missed the feet. I saw you washing 'em off with the hose. They're still sh--ty. Now trim 'em!"

Trimmer: "Aw, bulls--t!"

Inspector: "Naw! Hogs--t, and it don't belong in our cooler. Trim them d--n feet."

Enough of philosophy. On with the stories!

In the Beginning

The time was 1933, shortly after the election of Franklin Delano Roosevelt. The place was a government meat inspection office in a packing plant in Kansas City. It was hardly larger than a closet and sparsely furnished. It was occupied by two government veterinarians; Dr. Sullivan and Dr. Smith.

Even now I can see Dr. Sullivan in my mind's eye, though that is now dimmed by age. I remember Smith well, but then he was much younger. Sullivan told me how it was when Smith came to work for him. "He was a good kid."

It was hard to imagine Smith as young but I'm sure he once was, like we were once. When I knew him he was a fat, florid, middle aged man who seemed to have too many ducks to keep in a row.

Sullivan I met only after he had long since retired. It was not long after the war, World War Two, of course, when I had been too young to serve.

George Patrick Sullivan, named for the patron saints of England and Ireland, had been born near Boston well before the turn of the century. The good sisters of the orphanage had reared him and named him. He said he was Irish. That is, the sisters had told him he was of Irish parentage and it must have been true. He surely looked Irish. He was a large man, over six feet tall, very muscular, not fat at all, with hands that seemed too large even for him. When he was young his hair was thick and red. Freckles spangled his face. The only thing that didn't seem Irish about him was his presence of mind under adversity for if angered or annoyed he concealed it perfectly. Of that we will speak later.

Smith came from Texas. In fact, Wilber Smith came straight from Texas Agricultural and Mechanical College's School of Veterinary Medicine into Dr. Sullivan's inspection

10

office very early one morning, during the Roosevelt administration. He explained that having been examined, appointed and hired, he was now Dr. Sullivan's assistant. This suited Sullivan very well. Smith was used to handling cattle, although these pigs were a bit new to the young man from Texas. He knew they had lots of terrible teeth, rather like a small hippopotamus. In time he learned that generally they had a friendly or fearful disposition and seldom were dangerous.

"What can you do? Can you cut head glands?" asked Sullivan.

"Sure," said Smith.

"Have you got your equipment? Knife, hook and scabbard?"

"Yessir!"

"Let's see if that knife of yours is sharp. Hand it over here and let me have a look at it."

Smith took the knife out of a sort of briefcase he was carrying and started to hand it to Sullivan who immediately said, "Hold on there. First, when you hand someone a knife, point the handle at him with the blade pointed back at yourself and the edge up away from your wrist. That's the safe way.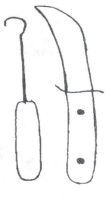
Furthermore, don't point out things with the knife point. Is your scabbard in there?"

"Yessir," said Smith, as he had been taught at Texas A&M, and by his father before that.

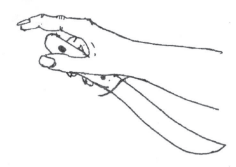

Smith must have learned this lesson very well for I remember how he would fold the knife back along his wrist and point with his index finger into a carcass to show us something he wanted us to see. It seemed silly. What could he hurt in a condemned carcass with the knife point and edge? However, I'm getting ahead of myself.

Sullivan took the knife, ran his fingernail lightly along the edge of the blade and then took a discarded envelope and cut it neatly into little slivers.

"Let's see your steel."

A steel is a round metal rod, tapered, often with grooves on one side, with a handle and hand guard, rather like a knife but used to sharpen knife blades. The hook is like an icepick with a recurved point.

"Yessir," said Smith and having gotten out his steel, hook, and scabbard from the bag, handed over the steel. Sullivan ran the blade along it gently, on both sides and then tried it again on the paper.

"Pretty sharp! Well, now, let's see how sharp you are this morning. We'll see how you do on cutting head glands."

Dr. Sullivan showed Smith where to cut and explained why the examination was conducted as it was.

Quickly he incised several lymph nodes into neat slivers that he turned over as he cut as though turning the pages of a book. He also cut the cheek muscles into thin sheets. The lymph nodes might reveal abscesses or tuberculosis; the muscle might reveal parasitic cysts. Either could be deadly to the person who ate the meat if it were poorly cooked.

The morning wore on. Smith did not mind work, but having been taught much about all sorts of bad things that he might find, became bored when nothing much showed up. At lunch time, he remarked upon the situation. Dr. Sullivan promptly told him that he should be glad when it was dull work.

Dr. Sullivan left for a few minutes to see how things were progressing in another part of the plant. The office seemed warm and close to Wilber Smith.

Suddenly he was sitting in an audience in a dark auditorium. On stage was a chorus of men in white clothes. They wore very large white hats and large silver badges. They were singing a tune he recognized as "Hail to the Chief", but the words were different:

"Congress has decreed that the packer is a Bad Guy,
Killing little animals to line his greedy pockets.
You will enforce all the Federal Regulations.
Raise Cain with him at every opportunity."

He could not catch all the words but it all sounded very noble and uplifting.

There in center stage, resplendent in his white uniform stood Dr. Sullivan. "Will Candidate Wilber Smith please come forward?" He got up from his seat and went up onto the stage. "Kneel. Do you solemnly swear to perform the sacraments of Retention and Condemnation, as prescribed in the Regulations, (The Manual of Procedures

13

was still some years in the future.) passing only product which is clean, wholesome, nutritious and sound; so help you Theodore Roosevelt and Upton Sinclair?"

"I do."

"Then rise and receive your official badge. Now you are an Inspector."

Bells were ringing, steam whistles blowing, and he was being lifted onto the shoulders of the other inspectors.

Suddenly he realized that he was being gripped and shaken by the shoulder. Dr. Sullivan had returned.

"Confound it Smith. Wake up. It's time to go back to work. First day on the job and you fall asleep. By the way, here is your badge.

"Tomorrow, be here at 5 o'clock, so you can go with Mr. Boswell on presanitation inspection. I may be with you too."

The morning came very early, but it was soon daylight since it was early summer.

Mr. Boswell had little to say except, "Come with me." He carried a flashlight, his hook, knife and scabbard, and most important a clipboard, with a form on it.

He said, "Good morning," to everyone he met and they would return a welcoming gesture. He looked at floors and ceilings. He looked at the bottoms of empty barrels in the tank room. The kill floor foreman and Smith followed him and watched carefully. He carried tags in his shirt pocket which said "US Rejected" in big letters. There was a big device in one corner that resembled an inverted cone. It was called a tripe washer, used to wash cattle stomachs so they could be used for food. He stopped and his flashlight lingered for several moments on some yellow material stuck to part of the machine. He said nothing, but took out a tag, wrote on it and tore it in two. One part he put back in his pocket and the other part had a hole with a string in it which he used to tie it on the tripe washer.

The foreman directed another man following to start cleaning the tripe washer. Boswell went on to examine other parts of the plant. He squatted down so as to look under machines and on other occasions would run his fingernails along under surfaces to determine if there was hidden dirt. "I'm underhanded," said Boswell with a sly grin. After he had found three or four places which were less than clean, he turned to the foreman and said, "I guess you aren't ready for me to inspect yet. Dr. Smith, let's go back to the office."

It had only been a few minutes and Boswell had not examined everything. The foreman and his helper muttered that Boswell was showing off. It might have been true for oftentimes he would just point out the dirt and it would be immediately cleaned. Every so often it would seem to them that he would be much harder on them than at other times, and they often thought it was unfair. Actually the inspection service wanted to have the plant people check up on themselves and the sanitation inspection was supposed to be a sampling of the kind of job the plant people were doing. This was a sort of new idea in the thirties and only after the war was it fully put into practice.

"They weren't ready, Doc."

"Oh." replied Dr. Sullivan looking up from his reading. They had been inside the office for only a few minutes when in came a man who said, "We are ready now, Doc."

He had been looking at Mr. Boswell, but it was Dr. Sullivan who answered, "Okay, we will be there in just a minute. Now I'm going to ask you to have a ladder so I can get up and check out the overhead rails. Then I think Mr. Boswell will take another look at the trolleys and head racks, after that, we might take a look in the boiler room..." His voice trailed off as though he were thinking about where else he might look.

The other man's expression darkened and became fearful. He was young and inexperienced and was far from home in a strange land, and found himself in a no man's land between his foreman and the inspector. He stammered that perhaps they weren't quite as ready as they could be and that he would check again to be sure that the kill floor was really ready.

"Do you mean to tell us," roared Dr. Sullivan, "that you came over here and you really were not sure you had everything cleaned up?" The possibility that Sullivan would be looking for dirt was terrifying for he was wise to the ways of packing house workers and knew just where to look for dirt from long years both as worker and as inspector. Furthermore he was imaginative. There was no telling just where he might look or what he might find.

The young man gulped, turned around, and quickly disappeared. He did not come back. After some minutes, the kill floor foreman came back and said, "I think we are ready."

Doctor Sullivan almost said, "You think you are ready. Why aren't you certain?" but instead said slowly and deliberately, "Okay. We will come and take a look."

So they went and looked, but if they saw anything, neither Sullivan nor Boswell said anything, although at one point Boswell pointed at a spot and a worker hurried to use the hose. Boswell looked at the tripe washer again. It was as clean as seemed possible in the days when galvanized black iron had not been replaced by stainless steel and exotic plastics. He cut the string, removed the tag, and placed it in his pocket.

So they went to work. The affair had taken almost an hour. One of the workers muttered that they might throw their weight around in this plant but it would take real courage to throw their weight around over in "Old Nasty". The trouble was that on occasion inspectors had done just that and received veiled threats. More often there came a call from high places to take it easy on them over at "Old Nasty." After all the plant property had been condemned and soon it would be shut down.

The days wore on. Summer seemed endless. The hot sun rose early, but not as early as Smith would have liked, for it seemed that he was always going to work in the dark. He had found a small boarding house not too far from the packing house district where a pleasant middle aged woman served breakfast and supper to her few boarders. Most of them also worked in the district and had similar hours. Some spoke little English. Smith had some speaking knowledge of Spanish but no understanding of Hungarian, Serbian, Russian, or Greek. So he could only listen and wonder at times. Mostly they seemed good hard working young men with families elsewhere. Most important to Smith, it was a respectable place. The landlady made that clear from the first day. There were to be no girl friends, no drinking, no gambling, no foolishness in her boarding house.

This suited most of the men; for she took in only single men. There was little they chose to do but eat and sleep there. One could read, and sometimes play cards, but not for money. On Sundays, Smith would ride the streetcar and see the town. He went to different churches, all Baptist, with an eye to seeing which had the prettiest girls before he settled down to just one.

Usually he would write his father and mother letters about his experiences in the plant or the city.

Sometimes he would work on Saturday. It was called overtime. When the packers needed to work on Saturday, the inspectors who then also had to work were paid at a higher rate. The workers were supposed to be paid more as well. Not everything worked quite as it was supposed to since this was during the depression. The new administration and congress in Washington had passed many new bills affecting labor. Some years would pass before things really changed.

Working on the kill floor at the plant with two other inspectors, and some thirty odd other workers, was actually an isolating experience. Smith was a Democrat. Something he had inherited along with his Baptist religion. Sullivan, a Boston Irishman, was also a Democrat. Sullivan, however, was far less sure about all the new legislation in Washington being able to bring the new kingdom to earth. His experience in the Army, and later in the Bureau of Animal Industry had led him to be suspicious of government. He had a generally low opinion of law and government in their ability to make bad people good.

There was so much noise in the packing plant that conversation was impossible. Hand signals were often used. To get someone's attention, one had to tap him gently on the shoulder, or if he were some distance away, you would throw a small scrap of meat at him, perhaps hitting him on the back with it. He would turn and face you and then you could wave at him. The noise eventually made Smith deaf. It had another effect. It made him talk louder than normal. It was a loud booming voice, easily heard above normal conversations in a crowd. Yet this was the way most packing house veterans talked. They didn't shout but they were loud. Wives, children, parents and fellow office workers would complain. Yet this condition pervaded industrial America for nearly another half century before it was discovered that loud noise caused deafness and measures began to be taken to protect workers' hearing.

The dust bowl had begun. Great clouds of dust carried away topsoil from overworked farms, and with it the hopes and livelihoods of those who lived and worked there. Emaciated cattle began to drift into the packing house in larger and larger numbers, and their condition seemed to worsen. At first there were only a few rather thin cows, then there were many, and then amongst these were first a few, and then many that were unfit to kill for food, or which were condemned for emaciation, or anasarca.

Smitty had seen "the holler belly" and sometimes "holler tail" in Texas long before he had gone to vet school, but this was new to him. One day while he was doing antemortem inspection two cows were dragged out of the cattle car. Neither could stand. Rather than subject them to being dragged farther, the weighmaster called Smith out to condemn them where they were. They were a pitiful sight. They had been trampled by the other cattle who could walk during the loading, shipping, and subsequent unloading. Only one could now hold her head up. There were raw bloody places on their hides where the skin had been scraped off. Along the bottom of their necks and bodies, the flesh was swollen and unnaturally hard, something the books called bottlejaw and ventral edema. Along the neck of one animal, he could see a sort of whipping motion as the blood coursed through the external jugular vein. This was called an external jugular pulse.

It took only a few moments to load the captive bolt pistol, all having agreed to kill and tank both the animals.

A captive bolt pistol is a very solidly constructed single shot pistol which accepts only a blank cartridge in its chamber. But instead of a rifled barrel containing a projectile, it has a plunger in a short tube with very heavy buffer spring and a flange which limits the travel of the "captive bolt" to a few inches. When the gun is fired the bolt plunges through the animal's skull into the brain instantly killing the animal.

Smith was writing out the antemortem condemned numbers from the condemned tags, to enter into the forms that he or Sullivan would fill out later. Briefly startled, he looked up only when the gun went off killing, the cow that sprawled out flat. At that moment the other cow, bawled and struggled to get up, but in vain. Tears now ran from her eyes. A moment later and she too was gone, skillfully shot by the weighmaster who was holding her by the horn with his left hand and with the pistol in his right hand.

Not having a holster, he shoved the gun beneath his belt and removed the bell and collar from the second cow. On the collar was the name "Bossy." It was a hard moment for Smith. His eyes dimmed; there was a lump in his throat, for he could see beyond the ruined cow, a ruined farm, its topsoil blown away, its ruined inhabitants, and the ruined dreams.

Something else happened one day. Another thin cow hung before him and he said to himself, as he inspected heads, "There's another bad one".

As he cut through a lymph node, his knife hit something gritty. He immediately stopped the line but only long enough to get the kill floor foreman's attention.

"Get Dr. Sullivan! This might be TB."

He completed the examination of the head, stamped it condemned, then dipped his knife and hook into the sterilizer.

Hardly had he finished the head inspection than he heard a shout from the eviscerator, the man who cut open the cow's belly and removed the internal organs.

"Hey, Doc, we've got a bad one, a real bad one!"

There wasn't enough time to stop and really examine the viscera. The other inspector who was working the viscera turned, spoke briefly with the foreman and then said, "Doc, here's a TB for you."

They pushed the carcass off onto another overhead rail, and put the viscera into another tub to await detailed examination. Soon Dr. Sullivan came. Meanwhile the killing of cattle continued as if nothing much had happened. Dr. Smith kept on cutting head cheeks and lymph nodes and Inspector Boswell continued to examine the internal organs as they were removed. As soon as there were a few carcasses ready to be put into the cooler, Dr. Smith would go over and look at them, paying particular attention to the kidneys and the lymph nodes inside the carcass. This was referred to as rail inspection. If something were wrong he would get the trim up man to remove it. These might be bruises or scars. Sometimes it might be something more serious, such as an abscess. An abscess was generally a round, hard hollow ball filled with pus, partially or completely imbedded inside muscle under considerable pressure. The pus would squirt out violently if the ball were cut into. Experienced inspectors and packing house workers generally learned to recognize abscesses and avoid opening them unless it was necessary. In any case, they would make considerable effort to cut around rather than into them. The abscesses were always removed.

Dr. Sullivan went over to Smith and said, "Go over there and get a good look at that TB carcass. You won't see many like that any more. I remember when there used to be a lot of them. I'll cut glands for you."

Smith stared at the carcass. It was very thin. Bones seemed to stick out at every corner. The ribs were clearly visible almost to the spine. The vertebrae also were visible up and down the back. There was an unnatural sheen to the carcass, as if it had been varnished with shellac. It was enough of itself to condemn the cow for emaciation or cachexia. But what he saw inside was the real climax. Inside the carcass were numerous pearly white objects, almost like clusters of pearls along the spine, and in the large lymph nodes in the pelvis. Later he would see more of the them in the viscera.

"Yep. It's pearly disease, the most advanced form of tuberculosis. We will have to examine most of the lymph nodes, check them off on our kill form, examine the hide for more lesions. And finally, most important, we must find out where she came from, who owned her before she came here," said Sullivan to Smith when he returned.

Fortunately, there was only a short time until the end of the work day so that after an hour more, the kill was over and they could return to the case of the tuberculosis infected cow. Sullivan reminded Smith that they had been lucky there had been only one. He remembered times when he had encountered several in a day and they all had to be examined thoroughly.

There was another bit of good luck as well as some bad luck. The weighmaster was able to tell who had sent in the cow. So Dr. Sullivan was able to inform the field service people. Not many days later another government veterinarian visited the farm and tested the herd of dairy cattle for tuberculosis. He and the farmer tested all of the cattle.

Testing for tuberculosis in those days consisted of injecting a drop of fluid called Koch's old tuberculin into the skin of the fold under the tail, a dirty but very necessary job. Three days later, the veterinarian came back to look at the tails, or rather caudal folds. Most of the tails looked just as they did the first time when he injected them, but some had swollen up as if in response to a bee sting. These he branded on the cheek with a letter "T", put a special tag in the ear, and wrote out shipping papers for the farmer to take the infected cows to market for direct shipment to slaughter. He also issued a quarantine for the farm.

First they had to catch the cattle. Since it was a dairy herd, used to being milked twice a day in a milking barn, it was not much of a problem. Now a herd of wild beef cattle, well, that was a different matter.

The quarantine would last for some months after the last infected cow had been removed to slaughter. No milk could be sold until the quarantine was lifted. Neither could any cattle be sold except to be sent directly to a government inspected packing house. There would have to be subsequent tests, separated by intervals of a month or more and the last two would have to reveal no more infected cattle. The vet called them clear or clean tests. Neither then nor now was there any sure cure for tuberculosis. There was quite a bit of paperwork. The farmer and the veterinarian talked about cows, TB, the dust bowl, etc. The farmer's wife brought them each a cup of coffee, and then offered them both cream. The veterinarian started to pour the cream. Suddenly his eyes glazed. His arm stopped in mid air as if suddenly paralyzed. He had suddenly remembered why he had come and what he had found.

The wife hastened to say, "That's store bought cream. Besides we pasteurize all of our milk."

"I'm very glad to hear that," he said now remembering that he had seen the equipment for pasteurization earlier. "It makes it far less likely that any of you have contracted bovine tuberculosis."

Fortunately, none of the persons on the farm had contracted tuberculosis. Also, Dr. Sullivan was able to teach Dr. Smith a great deal about bovine TB since the reactors, as the infected cows were termed, were sent to the packing house where Sullivan and Smith were working at the time. Some other farms were required to have their cattle tested as well since there had been a bull who had visited the infected farm and some cattle traded. A few other farms with infected cattle were found, quarantined, and the cattle also tested until they finally tested "clean." So for a while there were TB cattle to examine and write up papers on, both on the farm and in the meat packing establishments.

The kill floor foreman asked Dr. Sullivan, "Just what does bovine TB do to you if you eat it?"

"Nothing, if you cook it hot enough. But I don't think you would want to eat anything that looks like that."

"Lord. No! But we have both seen TB cattle that were reactors that came through and the vet couldn't find anything."

"Yep, and we cooked 'em. Right? Anyway, bovine TB usually attacks the bones of the spine in children, and causes deformed backs. Hunchback sometimes. Generally it comes from drinking the milk from TB infected cattle rather than eating the meat. It certainly is dangerous. Of course there is human TB which usually attacks the lungs. That too can be transmitted to cattle." Dr. Sullivan continued, "I have heard of a farmer with human tuberculosis passing it on to his cattle by coughing and spitting. Then when it gets in the cow it attacks the lungs."

There were other things that happened to cattle. One day Smith was looking at a cow on antemortem and the weighmaster said, "I think she's been hardwared."

"Yep, she sure does look like it, at that. Well, she's not bad enough to condemn out here. We'll see what she looks like when we get her inside," meaning that following her slaughter, the post mortem examination would show just what had happened to her.

"Put her in the suspect pen. We'll run her at the end. Here's your antemortem pen card and a suspect tag to put in her ear."

"O.K. Doc. Do you want her temperature?"

"Yep, might as well."

So at the end of the kill Dr. Smith found himself looking at the thin carcass of the old cow. There was nothing odd about the head, but when he got to the viscera, there was a very large flabby heart, and a big abscess between the heart and the stomach. He wondered whether to condemn the cow or not. After all, sometimes such things are a judgment call. He would find himself wondering whether this one was as bad as the one he had seen Dr. Sullivan condemn the week before. He would ask himself whether the problem was well localized, or could he see generalized effects. He noted a general edema beginning along the belly of the cow. He had seen it on the live animal, he thought, but wasn't certain. He did not find the piece of steel baling wire puncturing the stomach and extending into the heart sack which he suspected had caused the trouble. Most likely it had done its damage and then passed out of the cow's digestive system.

Unnoticed Dr. Sullivan had come up behind him and was watching. He could guess what was going through Smith's mind.

"When in doubt, throw it out! Get another opinion."

Dr. Sullivan took his knife and made a long slit along the belly. Clear fluid oozed out. He cut into a joint in the hind leg and more fluid gushed out. Then he cut into the joint on the opposite side.

"If you aren't sure, keep on digging. Usually something will turn up to convince you one way or another. Besides, you can always say to the man, 'Wait until I have finished my examination' until he's tired of looking at you. Just don't let him see you flip that coin. Don't forget, some of them, anybody would condemn, some nobody would condemn. Most of them will be in between and that's one of the things we are here for. Try to learn all you can and try to be consistent.

A moment passed, "Well, what do you think, Dr. Smith?"

Smith shook his head and said, "I can't say I'd want to eat it."

"Okay! I agree. You shouldn't put your signature where you wouldn't put your mouth. If you wouldn't eat it, don't pass it!"

Dr. Sullivan reached up and cut several large X marks on the carcass. Smith put the U.S. Inspected & Condemned brand on the carcass.

"Next month, perhaps, we'll rotate over to Old Nasty and you can get to see some really bad pathology."

The trim up man came, cut down the carcass, poured a sweet smelling green fluid over the meat, and finally wheeled the whole thing away. As he did so he remarked to Sullivan and Smith, "That's the kind of stuff Old Nasturtium likes; junk! Lampwick and Lightfoot used to buy those. They are gone, but the company still buys that kind of stuff."

The Story of Lampwick

It wouldn't have happened except for the remarkable powers of the coil Lampwick got from a wrecked Model A Ford. Lampwick, ("I burn my candle at both ends... it makes a lovely light.") whose real name was Garrigan, figured that with batteries and a switch he could really make creatures move. He got it all together with tape on a cane and it worked.

It worked, really worked. It would almost raise the dead.

A veritable Zeus - he strode amongst the hogs, casting lightning bolts at the slothful. You had to admit it was better than beating on them with sticks.

Some days when you looked out at the yard, and the men were driving cattle from the cattle cars out into the pens, it looked as if they were carpenters building a fence. If you looked closer you could see that instead of hammers they had sticks or canes, and they were driving cattle, not nails.

The shock stick wasn't new except in these parts.

Now in those days packing plants were several stories high and men drove the animals up to the top story where they were slaughtered and then as the animals were disassembled, the various parts of the slaughtered animals were dropped down chutes to lower departments where they were made into various products. There was usually a huge steam engine chugging away in the basement and belts ran all over the plant. The small electric motors we have now came in about this time, which was between the two World Wars.

There was a winding ramp with high sides that the animals were driven up. Like a square tunnel, with a gap at the top of the sides for light, it wound up and around the side of the building and over other smaller buildings. All of it was made of wood. Some of the wood must have been a bit rotten...but we'll get to that in a bit. Not all the animals were eager to climb the ramp, so Lampwick found the shocking stick useful. Hogs in particular liked to pile up. Hogs at the front of the column would lie down and dare those behind to walk over them.

Dave Lightfoot, the other man who worked the pens, preferred driving cattle, so that while they would trade places at times, usually it was Lightfoot working cattle and Lampwick working hogs. Lampwick was small and always wore his bowler derby. With his cane and his bent unlit cigarette glued to his lower lip; he attacked his job with grim determination. These, combined with his short nose, long upper lip, jet black eyes, and thinning gray and red hair, made him resemble a leprechaun.

Dave Lightfoot

Lightfoot was tall and thin and when he wanted to show off, he was quite graceful. "Light of foot and fancy free," he would say, doing a little jig, or perhaps a Paso doble, stepping gingerly out of the path of an enraged cow and giving her a great whack with his cane. Both men had a great command of the language, especially when it came to imaginative cursing.

One day when Bill, the kill floor foreman, complained that hogs were not getting to the shackling pen on time he was told that unless there was a way to get to the head of the column there was no quick way to clear up hog jams in the ramp. There needed to be a pathway on the outside of the ramp where a man could walk rather than being directly behind the animals.

It was a dangerous place, for if the animals came back, there was only a set of overhead braces to grab onto and get up out of the way. One could be trampled and killed. Of course, none of this was new, neither to Bill, nor to Lampwick.

"You and your damn shockstick. I can't see where we are getting on any better."

"Listen! The only hogs I can move are the ones at the back of the line. I can run 'em over the others if that's what you want. I can even pile 'em up over the wall if you want."

"Baloney! I'll bet you ten you can't even get the first hog over the wall, much less over the pile up."

"Okay. You have a bet!" said Lampwick.

It wasn't a good bet. At least from the standpoint of the hog who had a small chance to escape if he went over the wall on the backs of his comrades near the bottom of the ramp. After all, the wall was seven feet high, and the ramp rapidly rose up to four stories, some fifty feet high, even though some of it ran over other buildings. It would be a very long way down to the ground or the roof of some smaller building.

The morning wasn't very old when some of the hogs at the front of the line decided to lie down. Now Lampwick was at the other end of the line with his shockstick, so the rear end of the line continued to move forward over those lying down. Amidst squeals and grunts and curses the pile of struggling hogs got higher and larger but it did not quite reach the gap between the top of the wall and the roof. Then to Lampwick's surprise, it began to diminish. Ahead in the dim light of early dawn he could see the pile going down.

As usual, he thought, the pile had sorted itself out. Those on the bottom had struggled out. The din was deafening. Not that it made much difference to Lampwick who had been an artilleryman along with Lightfoot during the Great War. However, as he got closer he saw the pile was gone, and in fact, so was a part of the floor and wall. Hogs were slipping off into the blackness below. One could hear squeals, but the sound of hogs hitting the ground below was inaudible.

"Well, I'll be," said Lampwick when he discovered what had happened. He turned around and went back down the ramp. Outside, amidst the dead and dying hogs, stood Bill with a couple of the fellows from the kill floor.

"Do I win?" asked Lampwick. The answer seemed too obvious to even warrant the question.

Bill thought a minute. Kill floor foremen generally have great presence of mind, and great patience. Taking off his hat, scratching his head, he said, "Under ain't the same thing as over.

"Them hogs fell through the bottom of the floor, not over the top. That was just an act of God. You wouldn't want to say that you caused all this, now would you?"

The buyer and one of the plant owners, Atilla Farcas, were wandering around counting dead and injured hogs. "I could have been killed," muttered Lampwick. "Act of God...an act of the Devil, more likely." Soon after Dr. Sullivan and Inspector Reilley came and the sorting began. Dead hogs were tagged and sent to the tank house. Injured hogs were sent back to the suspect pen to be slaughtered at the end of the regular kill. There were a lot of injuries; bruises and broken bones to remove. The cooler was filled with pieces of hogs as well as whole carcasses. It was a very long day. Neither man tried to collect on his bet.

There was one good thing came out of it. The floor got fixed, and it had the separated walkway that Lampwick wanted.

It was a rough, bad, dangerous plant. Not everything was on the up and up. A lot went on that was best hidden. It was unfortunate, but it was one of those operations in which the minimum standards required by the inspectors became the maximum the plant could meet. It was the sort of plant that really justified the need of an inspection service.

It meant that the inspectors were always nipping at the heels of people who weren't quite doing their job. So that like the animals, the workers were driven by their employers, and the bosses by the inspectors. Little enthusiasm was shown for doing any job better than absolutely necessary. Part of this was because it was an old plant headed for eventual condemnation by the city for the expansion of the railroad yards, but the original owners had sold the plant and built a newer and better one.

It wasn't long after this that Lampwick came to a bad end. He had been visiting a speakeasy when two rival gangs settled a territorial dispute with gunfire. Poor Lampwick was killed in the disturbance. No one replaced him at the plant for a while.

The effect of the loss of his friend on Lightfoot was severe. Lampwick had done much of Lightfoot's thinking for him, and was his drinking, gambling, and wenching partner. While both of them had a sort of mean streak, Lampwick knew the difference between good and bad but simply often preferred the latter. Generally poor Lightfoot didn't know which was which. Lampwick often got Lightfoot into problems just for fun.

Sometimes Lightfoot would seem to be listening to the pigs as if they were talking to him. Now anyone who is acquainted with pigs knows that pigs talk and will talk to you if you care to listen, but they, like you, have better sense than to tell other people who would not understand.

Early one morning, the pigs failed to arrive at the shackling pen at the expected time. This was not unusual, since with Lampwick no longer in charge, and with Lightfoot running both cattle and hogs with a new employee as assistant, things just didn't go quite as smoothly. After a week or so the new man would be fired and another replacement would begin.

The bosses in the front office didn't seem to understand that it took years to learn about hog and cattle psychology. It wasn't just a matter of picking up anyone off the street. Lightfoot explained that the hogs didn't know the man and therefore they had no respect for him. It didn't seem odd to Lightfoot that each day the animals were new and different. He seemed to understand that there is a communal inherited knowledge amongst hogs (and cattle) that allows a hog, any hog, to recognize a hog man when he sees one and to disdain the novice.

After a while, Bill, the foreman, in company with Farkas, the part owner and stickman, went down to the pens to see what the problem was. They found Lightfoot standing with his hands folded leaning up against the outer fence. Hogs were everywhere. That is, they were everywhere they should not have been. They were in the parking lot, on the railroad track, and they were in the road on the way to downtown. Gates were wide open. This wasn't strange because if you want to drive animals into a pen the gates have to be open. This was one of those days when Lightfoot was working the whole thing by himself.

"What happened, Dave?"

"Lampwick told me he and his friends wanted to go down to Paddy's for just one more drink. I said 'Okay' but you have to be right back when your time comes. But then they all wanted to go and it didn't seem fair to play favorites so I said they could all go, just be sure to come back soon."

"This is no time for jokes, we gotta get those damned hogs back, before they get in trouble. I'll round up all the guys off the kill floor and try to see how many hogs we can round up and get back."

While it might seem to most of us that a hog couldn't be in much deeper trouble than to be in a packing plant, Bill meant that the hogs might cause trouble for the plant. For example, in the road a hog is a formidable obstacle. A collision between a Model A Ford and hog is only slightly worse for the hog than the car. Then too a hog at full speed can upend the unwary pedestrian. Finally, hogs brought back to the plant dead represent a loss - even though they can be put into tankage.

It was still dark and only the hogs close to the plant could be seen. In the distance a police siren sounded.

Farkas looked at Lightfoot and being a man with a strange sense of humor asked, "How did you know it was Lampwick? Could you recognize him? What did he have to say for himself?"

"Oh, it was Lampwick all right. No doubt about it. Same short turned up nose, same pointed ears, same red hair.

But he didn't have very much to say. He was in a hurry. You know how he is. No patience at all. You always did say he would come back as a pig."

It was true. Farkas had said as much. Now he was no longer sure that Lightfoot was joking and the hair began to feel sort of creepy on the back of his neck. He stared at Lightfoot.

"Really now, I did the right thing by him, didn't I? Letting him go, I mean. Besides you couldn't just cut his throat like any other old pig, could you?"

It wasn't beyond the limits of Farkas' imagination to believe that perhaps Lampwick could return as a pig. In his youth in Transylvania he had heard of many strange things, and to his regret, he had done many bad things during the Balkan wars. At times they troubled him. There were circumstances under which he could cut a friend's throat and consider it an act of kindness, even if the church would condemn him for it.

Shortly Bill returned with a dozen or so of the men from the kill floor. He stopped a moment at the little office at the weighing pen where Lightfoot and Farkas were still engaged in conversation. Noticing that Lampwick's derby hat still hung from a nail, he grabbed it and a cane, and the three of them went after the escaped hogs. Meanwhile more men had arrived and had gone down the tracks to bring back those hogs that had gone in that direction.

"He said he was going down to Paddy's," said Lightfoot.

"That's as good a place to start as anywhere," replied Bill.

As they walked, they encountered hogs coming back towards the plant. Other people not connected with the plant had helped herd the hogs back. A few hogs never were seen again, having been taken on the basis of finders keepers.

Soon they arrived at the vicinity of Paddy's Grill (and speakeasy). Two police cars had pulled up. There were several policemen walking cautiously about with drawn revolvers. One of them however was carrying a 1928 Thompson with the drum magazine. In fact, he had been carrying it in the squad car for over a year since it had been confiscated from some bank robbers and he had never had any excuse to use it.

Behind the building, headlights of police cars shown brightly upon barrels and garbage cans. Amongst them there was a commotion. Bill got a brief glimpse of a red pig running from one hiding spot to another.

"Come on out, Lampwick. You don't want to stay a pig forever, do you?" shouted Bill, perhaps partly to humor Lightfoot who was close behind.

Then the pig made a break for freedom.

No better excuse existed. The trusty Tommy gun barked and the pig was struck several times. Death was quick.

Bill was first to reach the dead pig. Bill rolled him over from his side onto his belly and propped his head up on a nearby loose curbstone. Realizing he was wearing Lampwick's hat, he took it off and put it on the pig's head. He took out the cigarette he had been smoking and put it in the pig's mouth. The effect was uncanny. The pig really did resemble old Lampwick, bent cigarette, bowler derby and all.

Lightfoot said, "Good-bye old friend."

Farkas crossed himself and said, "We'll give him a good send off."

Soon another man came with a wheelbarrow. They placed poor Lampwick, the pig, in it and took him back to the plant and thence to the tank hose.

Lightfoot was sent home. Evidently he had been working too hard, or perhaps something was wrong with the stuff he had been drinking, or perhaps it was something the girls at Aunt Maggie's had given him. Illegal alcohol during prohibition sometimes contained lead and the girls in institutions of ill repute sometimes had syphilis. Either could affect the mind adversely.

"So that was the story of Lightfoot and Lampwick," concluded Mr. Boswell, the inspector.

"Poor old Lightfoot wound up at the state mental hospital," stated Dr. Sullivan. "That's been a few years ago."

"Yeah, but Farkas is still there at Old Nasty and he is a bit odd. They say he drinks blood."

"He says he drinks blood," said Sullivan, "but I never saw him do it. Says it tastes just like tomato juice."

"You could sure catch something bad that way...like Bang's disease," exclaimed Smith.

"Yep, but he's probably already had it."

That night Smith went to sleep wondering about what it would be like to be a pig in the packing house, whether being hoisted up by the hind leg really put them to sleep. He doubted it. He wondered how badly the knife hurt. He knew that bleeding to death was nearly painless. But the cut was something else. Many years later hogs would be gassed with carbon dioxide, or shocked with electricity into insensibility.

Suddenly he was in a huge room. It was a packing house. No, it wasn't Old Nasty. It wasn't even K.C. Packing. It was far cleaner and better than anything he had yet seen.

Around him were his fellow inspectors and the plant workers he knew and worked with. Again they were resplendent in white uniforms. In front of him was a huge tank. Leading to it was a sort of chute with a huge machine called a hasher that could be best described as a giant kitchen sink disposal unit. Into it could be dropped an entire condemned carcass which it would grind up into small fragments in a few seconds. He had heard of them but never seen one.

Also on the chute was a coffin, draped with an American flag. Inside was Lampwick.

Lightfoot began the service reading from the Bible.

"Man that is born of woman hath but a short time to live..."

Then Dr. Sullivan read from the Regulations:

"And the body which is corruptible shall be tanked according to the Regulations, but the soul which is incorruptible, having been trimmed, inspected, and passed shall be stamped with the marks of inspection and shall hereafter be placed in the great Chill Cooler in the Sky."

"Let us now commit the body to the tank."

Bells were ringing, whistles were blowing, the coffin slid down into the hasher and disappeared.

Suddenly, it was five AM. The alarm clock was ringing and it was time to get up and go to work.

The Girl Without Hands

"The worst accident I ever saw?" answered Dr. Sullivan.

The question was in response to a question by Dr. Smith. It had been raised by an accident in the cattle pens. Smith had been conducting ante mortem inspection when the cattle had suddenly panicked, turned about and stampeded down the alleyway only to pile up at the end of the lane. Smith and Jones, who was the pen master, had leaped out of the way just in time, or they might have been trampled to death. Several of the animals were killed, and several were injured. Both men were experienced with livestock and generally knew when the animals were edgy, or in a dangerous mood. Jones was a kind and reliable man who understood cattle. It was just one of those things that happens sometimes and has to be watched for, Jones had explained.

Dr. Sullivan thought for a few moments and then said, "I saw lots of bad things during the war. But the only accident that happened to anyone I knew personally was when Lt. Fredrick got kicked by Old Reliable. I'd call it an accident, since she was trying to kick a dog barking at her and hit Freddy. Broke both legs just below the knee. It was pretty bad. He had gone through the entire war with hardly a scratch and then this happened the day after Armistice day. Took him months to get on his feet again. Even then he had to use a cane and wear a brace for some years afterward.

"Come to think of it I know of a worse accident but I didn't see it happen. But it's another story."

"How is that?"

"Just after I finished up at the Indiana Veterinary College, I went to work for the government at a small packing house near Boston called Levy and McGinnis. I had worked in the area and gone to night school to get into vet

38

school. It was rough, working the kill floor as assistant foreman during the day and taking college courses at night. Fortunately, the requirements for getting in then weren't as stiff as they are now. But then I was a man who didn't have to be told twice. I learn pretty easy and I don't need much sleep.

"Ever since I began in the packing house I had wanted to be a vet. Now I was one. I know that vets can set up a horse practice, or even work on cats and dogs here in the city and make money...or they could when there is no depression, but I wanted to be a packing house government vet. I wanted to be somebody important.

"I guess I was lucky because the school closed soon after I graduated. Now nearly all of the old city schools are gone and veterinary schools are almost all connected to land grant agricultural colleges.

"I was standing there one Friday morning while the plant people all filed in to work. Ed McGinnis and Joe Malone were with me. I had just come back from doing ante mortem inspection on some lambs. It would be a short day. I would go to another plant when the kill was over for the day. Ed had become the plant director after his brother Douglas died a few years before.

"Joe, the kill floor foreman asked what kind of animals we had for the day. I told him, 'Succulent young stuff, lambs and young pigs.' "

"Now that's what I call succulent young stuff," remarked Ed with a chuckle as the ladies who worked in the sausage kitchen filed in. First came Mrs. McGinnis, walking with her daughter. Then came two younger girls, both cousins, and finally Mrs. O'Toole. Mrs. McGinnis was the widow of Douglas, the other son of the founder of the firm, old Jock McGinnis. Shortly after the founding of the firm, Jock McGinnis had arranged a partnership with Saul Levy. He observed that a Scotsman and a Jew should make money, one to make it and the other to keep it. It was a private joke, but it seemed to come true, for the business thrived. The money earned was put back into the business and it grew.

Mrs. McGinnis did not approve of Mrs. O'Toole, better known as Bubbles O'Toole. It was a stage name, rather than a given name, for Bubbles had been an entertainer in her younger days. She bulged in all the right places, only now with middle age approaching, she had a little too much in all the right places, even for Minsky's. Minsky's was a well known burlesque theater during this time.

This did not bother Isadore a bit. Isadore was her protector, as well as her boyfriend. Both had been married before. Isadore was the oldest of Saul Levy's three sons.

Izzy, as he was called, had married Golda shortly after he finished at Columbia. Unfortunately, Golda had visions of reshaping Izzy into an intellectual gentleman. Izzy and Golda did not get along and it became evident that the family who had promoted and arranged the marriage had miscalculated. Golda liked the concert and the opera; not for the music, but to be seen in the right places. Izzy liked Atlantic City, race tracks, poker, and jazz. What was worse, he liked the people; gamblers, race horse jockeys, honky tonk jazz musicians, and the lady entertainers at Minsky's.

Izzy was happy when the world was going meshuga, i.e. nuts, with a pencil behind his ear, a cigar in one hand and a telephone in the other. Since he was the oldest of the three brothers; the sons of Saul Levy, he had to take over the financial and sales part of the business. Golda would scold him asking why wasn't he like his brother the doctor or his brother the lawyer...respectable. It wasn't long before Golda left him. It was somewhat longer before there was a divorce settlement. Golda had to have one before she could remarry, and Isadore willingly gave it to her.

Bubbles had married a sailor who had been lost at sea. She was the widow of a serviceman, which carried few advantages other than a small pension. She had gone on stage to use her physical assets (which were ample) to support an infant daughter. The daughter lived away at school, and no one at the plant had ever seen her.

Joe was the young man who had taken over most of the kill floor operations when Stewart had gone off to the war. Somehow, he, like Isadore, had been declared essential to the war effort. Ed had taken over most of the duties of his brother, Douglas. In a way, it was said that Douglas, as well as his son, Stewart, was a victim of the Great War. A week after learning of his son's being killed in action, Douglas had died of influenza.

Joe had once been the object of Miss McGinnis' affections, but he had suddenly changed his mind and married her young cousin, Susan. There were whispers about them for it was not long after the wedding (though it was long enough) that she presented him with a baby daughter. Joe was not as able as Stewart had been, nor for that matter, neither was Ed. It was an unfair judgment for both men had too much to do. Two could not possibly do what four of them had been doing. So much of what they did was less than adequate and the business suffered.

Monday was the day that Ed and Joe had chosen to move the old number one boiler. It was old, damaged, and obsolete, and it was in the way of another room they wanted to enlarge. First they had to place screw jacks at each end of the boiler. Then they had to raise each end. Only a few turns could be made at one end before they would have to crawl to the other jack and in turn raise the other end so that the boiler remained nearly level. Unfortunately, there was a forest of pipes under the boiler. There was the need for undoing some pipe fittings first. Timbers, actually four inch wood posts, had to be carried in and used to brace the boiler

41

when they moved the jack. It was an elaborate operation that they should not have attempted without more help.

Meanwhile, the others in the plant labored at customary duties. The women were making and stuffing sausage in the kitchen. Other men were cleaning between kill floor operations. Ed and Joe began the long and difficult job of moving the boiler. They had just gotten the other end partially raised, when there was a loud cracking, grinding noise and they could feel the boiler just above them move to one side. They had not yet placed enough of the timbers for their own safety.

It turned out that they had placed the first jack over an old terra cotta drain and the base of the screw jack had broken through the floor and was tilted dangerously. If more of the floor gave way, or the small pipe against which the boiler rested gave way, then the boiler would fall on them, crushing them to death.

At the same moment the whistle for lunch break sounded, and in a few moments the ladies came out of the kitchen going towards the lunch room. First on the scene was Miss Martha McGinnis. She immediately understood the situation and at the men's request, began to lift and hand the heavy timbers through the spaces between the pipes to the men trapped inside. Once the boiler was braced, the jack could be moved, the men could get out, and plan what to do next.

She could see Joe and Ed between the forest of pipes. There were only a few of the timbers left. She picked up one of the timbers and shoved it between the pipes where Joe could reach it and then he and Ed placed it where it was needed. Soon she had given them several of the bracing timbers.

There was one final timber to place, one which would block the boiler from sliding. By this time others had gathered and were there to help. She had to slide one last timber between two pipes on the other side and she could see where it had to fit over another pipe. She had to reach in as far as she could pushing with both hands and all her strength. It was in a place the men could not reach.

Perhaps it was the vibration from a street car outside, perhaps it was the steam whistle going off again, but at that same instant the jack crunched through the concrete dropping the boiler several inches and crushing several of the pipes together.

The men escaped injury because the timbers had been well placed, especially the last one. If the girl screamed, no one heard her, and when they realized that she was hurt, they found that she had lost consciousness. She had been struck on the head when the collapse occurred.

It took several minutes to release her. The jack had to be repositioned with a steel plate under it.

When they got her out it was necessary to place tourniquets on both of her arms. The bones from her wrists to well above her elbows had been badly crushed giving rise to multiple compound fractures that bled profusely. The doctor arrived promptly, and after giving her a shot of morphine, had her taken as quickly as possible to the hospital.

In the operating room the surgeons were talking.

"That's very bad," the older surgeon said, "Multiple penetrating fractures, contamination of the wounds," and as he released the first tourniquet. "No circulation below the level of the elbow. It will have to come off, in fact they both will."

"What kind of life can she have like that?" said the younger man, who had not been a military surgeon. It was more a declaration than a question.

"She will be able to do most of the really important things, to marry, have children, and the rest."

"Medically, I must agree, I hate to see beautiful things broken. Amputation, upper third of the humerus, left arm."

A few minutes later, "Amputation, upper third of the humerus, right arm. Note, save them."

The severed limbs were to be kept for a while for the morticians, so that should the victim die, she might be buried intact.

The evening newspaper said;

"Heroic Girl Loses Both Arms, Saves Both Lives."
Young Martha McGinnis lies near death
tonight having been severely injured, saving the
lives of her uncle and of her former sweetheart.

There were photographs of the three of them, taken in happier times when they were younger, and a note...story on page six.

Probably those on the scene when the accident took place did not realize what had happened. Except for Mrs. McGinnis, everyone else had gone back to work. There was a calmness about the place, a sort of gloom. In the evening, then, they would find out how Martha was, ask about how soon she could return to work. Certainly no one wanted to imagine how bad it really was.

At the hospital they found that no one could see her. Her mother had gone home in seclusion. They had found out that she would not be coming back to work any time soon.

For several days she hung between life and death. There was a question about kidney function. Fortunately she had been young and strong. There was the good fortune that there was no infection. Nearly a week passed before they let anyone see her, other than her mother. Mostly she slept under sedation.

Lying in bed, Martha thought about her accident and her injuries; the how and why of it. There seemed an inevitability to it that made it seem inescapable. Why not someone else? Susan or any of the other ladies would have panicked and would eventually have run to get the men to help. Then it would have been too late. Bubbles might have tried but would never have gotten the timbers placed where they would have been effective. Neither could she think of any one who could have borne her affliction any better. She could not wish it on anyone else. She accepted her injuries as the price to be paid for the lives of the two men. She would have been willing to give her life for theirs, but this sort of living death...why?

Joe and Susan were perhaps the first to see her in the hospital. They each kissed her on the cheek. Susan burst into tears, confessing that she and Joe had been sinners. Martha stirred briefly and said, "You are forgiven, but you must both promise me that you will always be loyal to each other, and love one another so long as you live."

They agreed to this and to their credit they kept this promise.

Eventually they took her home. Her mother took care of her as best she could, which took up most of her time. It made it nearly impossible to work at the plant any longer. There would have to be hired help if Mrs. McGinnis were to go back to work. She really had no idea how to cope with her problems. Too much had happened to her. She spent a large part of her time wistfully thinking of a happier past and wondering why God was so angry with her. Surely she had sinned somehow to have had so much trouble.

The kill floor crew had a short day that day and the inspectors had left for other duties. Sullivan did not learn of the accident until the next day.

Dr. Sullivan wondered whether he should visit Mrs. McGinnis. There wasn't any help he could offer. Sympathy wasn't worth much. They already had enough of that. Then

45

one day about lunch time he was told, "There's a young man here to see you. Says his name is Fredrick, John Fredrick."

"Well I see you did it Sergeant Sullivan. Good for you," said his young friend. "I just recently finished my last and I hope final bar exam. I'm going to take a little time off before going into a partnership downtown. I have some connections and opportunities."

They talked for as long as the lunch break permitted, promising to meet and talk at great length later, which they were able to do the following evening over a good meal.

Conversation turned to the war and those they had known; of what had happened to each of them since, of those who had not been so lucky as to live through it, and of those who had been wounded and their fate. Inevitably Lt. Stewart McGinnis was mentioned.

Dr. Sullivan had heard about plans to sell the plant through the usual rumors that floated about the plant. It occurred to them that perhaps the two of them should go visit Mrs. McGinnis. Mr. Fredrick to carry his sympathy to her in the death of her son, and perhaps secondarily to offer legal help, if needed.

Mrs. McGinnis had met them at the door and led them into the parlor. Then she went to get her daughter and to brew tea for them. While she was gone and they were awaiting Martha, they looked at the photographs on the wall. First there was a picture of Jock and his wife with their young children. There was another of Saul Levy and his family.

Dr. Sullivan pointed to one of the small children in the picture and said, "That's Isadore, who now runs the sales and financial end of the business. Since Douglas died, Ed has done most of the construction, maintenance, and physical planning for the plant. I knew Douglas from the time when I was a kill floor employee in this area. I knew his boy, Stewart, too."

There was another picture of a younger man, Stewart, in his Army uniform.

"That certainly looks like him," remarked John, speaking very softly as if not wanting to break the silence. "Damned pity."

"Yes, I really think it killed his father too. You know he caught influenza and died almost within the same week that his son was killed in action."

"Terrible stuff, he was at work on Friday, and dead Monday night, and he was a relatively young strong man, only forty eight."

Dr. Sullivan had lingered to look at the photograph a bit longer, while John Fredrick had moved around the wall to see more. They had both known Lt. Stewart McGinnis, and considered him a friend. It had troubled John that he had put off seeing the McGinnis family. Dealing with the death of a friend was hard. Now that another tragedy had struck and a request for help had come, he could delay no longer.

Being partially deaf, and having turned his face to the wall, Dr. Sullivan had not seen Miss McGinnis enter. He was aroused from his deep thoughts about the persons in the old photographs by a commotion behind him. He turned to see a nightmarish scene.

The young lady had come in so quietly that neither man had been immediately aware of her presence. She was wearing her father's old dressing gown which was several sizes too large, tied tightly at the waist. Before she could even say hello, she had tripped on the rug, for she was still very weak, even though it had been nearly a month since the accident.

John had reached for her to break her fall, but had only caught her collar, and although that had kept her from falling on her face, she had landed in a heap, and the dressing gown had been pulled from her shoulders.

She was sitting with her legs akimbo. First, forgetting her handicap, she tried to reach down and pull the robe back up over her, when that failed, she tried to cover her exposed breasts with her hands, which she could not do either. What happened was that she waved her still bandaged stumps about, tossing her breasts.

Sullivan wanted to help her to her feet, but there seemed now no way to take hold of her. If he offered her his hand she could not take it in hers. It seemed impossible to grasp her by either arm for neither was any longer than the breadth of his hand. He knew by the bandages that it would have been painful.

Behind her sat John who had also fallen. He had not fully recovered from his accident with the mule and was not agile.

Somehow she seemed shrunken, and much older than she had been before the accident. She had certainly lost weight. The color had gone from her cheeks. Her tears and her hair covered her bosom. There was a bruise on her forehead, the result of a previous fall. Without arms and hands, a simple fall could be very bad.

She whispered, "I'm helpless, useless, I can't even go to the bathroom by myself. I have to be taken care of like a baby. I wish I were dead."

Sullivan wanted to say, "Hush, child it can't be that bad. You will be all right." In his heart he felt that perhaps it was that bad and so he said nothing, and tried not to stare at her.

John took hold of the robe and pulled it up around her shoulders covering her again. Then still sitting behind her asked, "Let me help you up?" Then, without waiting for any answer, placed his hands firmly about her waist and lifted her up. Then, using his cane he got himself up.

They were all standing when her mother returned from the kitchen with a tray containing tea cups and the teapot, which she placed on the table.

They did not tell her mother what had just happened. That she had fallen before was evident from the bruise on her forehead.

Mrs. McGinnis began, "We need the services of a lawyer. Stewart mentioned you in his letters. Evidently he thought highly of you."

"Isadore Levy wants to sell the business. Considering what has happened to us, it is probably the best thing to do. We will need the money. We need someone to help determine just what part of the business we own."

Jock and his sons after him had put much of their earnings from the plant back into the business. For that matter, so had old Saul Levy, but his heirs had tended to put their money into other things. These other ventures were often successful, but had reduced their share of the business. The womenfolk had not kept themselves well informed of the financial part of the business.

This meant pouring over old papers, deeds, and stock transactions dating back over the years.

Fortunately John Fredrick and Isadore Levy got along rather well. Isadore wanted to get his money out soon. There was a real estate boom just beginning in Florida and he wanted a quick settlement. In the meantime, there was the need to find a buyer for the business. The two of them might have made successful business partners, but it was not to be. They would each go in different directions once the business was sold.

One day John came to visit the home on business. It seemed that it was a three-way thing- visit the courthouse, look at records; visit the plant office, look at records there; go back to the McGinnis home and look at records there.

John looked at Martha and then said to her mother, "She needs to get out for a while. Staying shut in isn't good. She needs things to do, fresh air and sunlight."

Martha might have argued had she disagreed. Being caged inside was no fun. She had not been out of the house for a month.

"How about letting me take the two of you to church this Sunday? I'll come pick you up in a taxicab and we can go from here."

They agreed. As soon as John had gone Martha

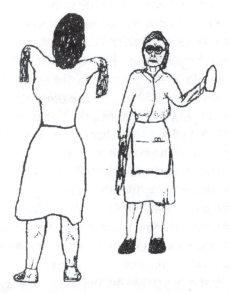

began, "What can I wear? I can't be seen like this."

She raised her stumps with the short sleeves of her dress dangling from them and said, "My arms haven't even been ordered."

She meant that the doctors had insisted that it would be some months before her stumps had shrunk properly so that she could be fitted with artificial arms.

"We'll get you a cape to hide in," said her mother. It wasn't exactly what she had meant to say, but it came out that way and it hurt her daughter's feelings, and Martha said nothing.

Sunday came and church was less eventful than they had expected. First, they had gone to a different church uptown. If anyone stared at Martha, she was unaware of it.

The next week they went to the family church they had always gone to in past years. Friends tried not to stare. Others tried to act as if nothing at all had happened to her. They tried to make her feel at home. Perhaps to some degree they succeeded. It was her mother who seemed to suffer most. John noticed her tears and asked if anything was wrong.

She answered softly, "Too many ghosts, too many memories. I was married here. My children were baptized here." She did not need to add that her husband's funeral had been there as well.

It was then that John decided that something more diverting than church was needed; something to lift their spirits. But what?

He looked up. There was a billboard. The circus was coming to town. Perhaps that would be a start. It would take some planning, but perhaps it could be done. It would be in town next month.

One day Martha asked John to take her to the doctor's office for her appointment. It was necessary to check to see whether she would be ready for fitting with artificial limbs. The answer was not yet.

In the drug store, near the doctor's office they stopped to share a soda. She could drink through a straw with little or no help and not attract any attention. As they finished, two small boys came up to the table and asked Martha, "Are you Joan Wisent?"

"No."

"Are you one of the Lee sisters?"

"No, who are they?"

"Aren't you with the circus?" asked one of the boys.

"Why, no," she said, not understanding the implication of the questions.

John, however did and blushed, and thinking quickly said, "No, but she's just as great as they are. But we're busy now. We'll see you there. You are coming to see the circus?"

They said, "Yes." Suddenly seeing their streetcar they turned and ran for it.

"Time for your appointment," said John, getting up.

It seemed that the surgeon was someone John had known from the war. He suggested that Martha should have some regular exercise and some sort of goals. It seemed that they both knew of a woman who taught ballet and modern dance to a few private students. It might be very good for her to learn better balance, to avoid falls and to gain confidence.

So it was arranged that she should start taking dance. It seemed strange to her but it made sense.

Just as they were getting ready to leave, the surgeon said, "I'm having a small get-together tomorrow night. I would like both of you to come."

So Martha and John found themselves at a small party in the home of the physician. There were perhaps a dozen persons. It seemed that they were all young men who were war veterans. Even the other woman who was there was a nurse who had served in the war. Each seemed to have some sort of injury. Some of them were already John's friends.

There was a piano. Martha thought about how she used to play. It was just as well that they sold the piano after Douglas and Stewart died. Anyway she preferred to sing and she could still do that.

So they sang the old songs; It's a Long Way to Tipperary, Pack Up Your Troubles in Your Old Kit Bag, Roses of Picardy, and the others.

The young man playing the piano had no sheet music to read and he was wearing dark glasses. He was blind, the result of a flash explosion in a ship turret that had singed his corneas causing them to develop white scars through which he could only distinguish light and dark.

Someday a medical miracle called a corneal transplant would let him see his grandchildren and his wife of some thirty years, but in the early twenties it was as remote as space flight.

There were other injuries represented, only a few as obvious as a missing limb, or blindness. The man at the piano finally stopped, turned, and spoke to Martha, "I love your voice. It's beautiful," he said softly. "May I hold your hand?"

"No! You can't!" she exclaimed in alarm. Then, realizing she had spoken harshly, she added, "But you may touch my cheek."

Evidently he could locate sounds well for he reached out with both hands and felt her cheeks, "I have to see faces with my hands. Then I can visualize faces. Of course, I must ask what color your hair is and whether your eyes are blue or brown. What has really been hard is learning Braille."

By this time he had examined her face, his hands passed quickly across her shoulders, and started down her arms, remarking, "You must be very pretty... Oh no. I'm sorry. I didn't know."

"Now you -- understand," she stammered. She had almost said, "Now you see," Tears welled up in her eyes, and for a moment she could say nothing. Blindness seemed so terrible, something she really could not endure, something much worse than what had happened to her.

"Are you the woman who saved those two men in the meat factory last winter?"

"Yes," she answered in a very small voice.

"You must be very brave," he declared.

"At the time, there was no time to think about it. I didn't know what could happen to me, only what could happen to them."

"Did you know them?"

"Of course. One was my uncle." She did not add that she had once loved Joe. "I need to be brave, but I'm not. I wish I were."

"What you say sounds so familiar. Men do brave things because they have not counted the cost, or perhaps, the bill of consequences will be delivered to someone else."

Then came the day to see the circus. There was an elderly gentleman in a large cape standing to the side. Evidently John knew him for he spoke to him calling him Mr. Trippe.

Mr. Trippe introduced himself to Miss McGinnis. He looked at her for a moment and noted. "You are one of us, except," he hesitated a moment before continuing; "Only I was born like this, and you appear to, ah, have become this way. I'm sorry. How do I know? By the shoes you are wearing and the way you stand."

John had arranged it and had told him about Martha's plight earlier, but he need not have.

Mr. Trippe continued, "Forgive me for sitting in the presence of a lady, but my feet are my hands and if I stand on them I can't do other things with them." Then he sat down and kicked off his shoes, or rather gaiters, which lacked shoelaces. His socks had holes in them so that his toes were exposed. The effect being like gloves with the fingers cut off.

"Let me show you something. Here is my business card."

So having said this, he reached into his coat and took out a small folder like a wallet. Using both feet as nimbly as most people use their hands, he took out a single card and held it between his toes. Then he took out his fountain pen from another pocket, removed the cap, and, holding the card in one foot, signed the card with a flourish after hesitating a moment, gave John the card.

Martha was incredulous. She asked, "Can I learn to do that?"

"Yes, but it will be hard. It won't be as easy for you as it was for someone like me who was born this way."

"The accident has made me illiterate. I can't write, of course, and I can't even pick up a book, much less turn the pages."

"Well, how well do you see?" he asked.

"Quite well."

"Try putting the book on the floor and turn the pages with your toes. It will be good practice and will help you pass the time. I don't suppose you have a friend with a bicycle whom you can ride with...as I used to do with Mr. Bowen."

The conversation went on for some time with John just listening. One part interested him quite a bit.

Mr. Trippe had said, "You know. I once met a brain surgeon named Wilder Penfield who wants to putter with my brain."

"Why on earth?" said Martha and John, almost in unison.

"Well, Dr. Penfield says that on the side of the brain is a sort of picture of a little man, a sort of homunculus shaped area controlling the motor functions of the body, or rather a sort of map or image. But it is distorted. It has a huge thumb almost as big as its leg. So he wants to know whether I have a little man on the side of my brain like every one else or does it have an enormous big toe instead of a thumb, and perhaps no arms.

"While I am interested, it is an honor I would prefer to pass by."

The conversation went on. Finally Martha asked, "Could I get a job in the circus?"

"You are very pretty," he answered, "and I feel sure you could but I must add that just being handicapped is not enough. What interests people is not that terrible things happen to people, but what they can do despite their handicaps."

The months went by. Martha enjoyed her ballet training. Color came back into her cheeks. But there was something else. She had become very fond of John but she had no way of knowing how he felt about her. He was always proper, cheerful, and correct. Someday, the legal work would be over. The plant would be sold. John's legal career would take him out of her life and she would be on her own.

Still, John seemed to visit more often than business strictly required. If her mother noticed this, she said nothing about it. She had heard that John had inherited a great deal of money, and really didn't need to work. He might have gone into politics, but lacked the interest. Meanwhile he studied business and corporate law.

The artificial arms were finally ordered. When they came, they were a failure. If she stood still, they filled the sleeves and made her look normal. They seemed very heavy, and they seemed to want to swing about in an unnatural way. She had them put on a shelf in the closet and she continued to wear her cape. If John asked her out for an evening's entertainment, it was nearly always as one of a group. Sometimes her Mother went with her Uncle Ed and his wife. Sometimes it was with Joe and Susan. There was always the need for another woman lest she need to go to the bathroom. It was only when they were in transit that they were really alone.

One evening as they walked home, they talked about their individual plans. He explained how he was orphaned when the Titanic sank, killing his parents. How he had lived with his grandmother, then gone off to prep school, then college. Harvard, of course. Then the war had come, and as a first year law student he had gone to OCS.

His grandmother had died during the war. Then he was alone, except for the two old family retainers who lived in the old house with him. Oh, and there was an old dog.

Mrs. McGinnis had entrusted the house key along with her daughter to John that evening so he could let her in after Mrs. McGinnis had gone to bed. They let themselves into the alcove. At that moment, Martha stumbled. This time, John caught her, and even after she had regained her balance, continued to hold her.

He was a little surprised at how strong and muscular she seemed. It was so pleasant just to hold her there in the dim light of the atrium. He put his cheek next to hers. There was a clean scrubbed smell about her rather than perfume. Her bosom pressed against his chest. He could feel her heart beat faster, like his.

"Please, I'm still a good girl," she whispered.

"Yes, so am I," he said. It seemed funny to say it that way. But it was true and they both began to laugh.

"We'll wake your mother," cautioned John.

"She will have to undress me anyway."

They both laughed again for the same improper thought had occurred to both of them at the same time.

More time passed. The details of the sale of the packing house all fell into place. Isadore and the others liked the deal. The employees, for the most part, would have the chance to continue in new premises which would be built later. The plant would continue to function for some years in the same location while the new owners, who were well backed, moved operations into the more modern plant soon to be under construction.

The inevitable day came when it was all over and done. There was the matter of the legal fees. For himself, John had charged nothing. There was the matter of costs involved which included copies of court records, stenography fees, and other matters.

But he took Martha out one more time. He took her to the opera, and this time, again standing alone in the atrium with her, kissed her hard on the lips and asked, "Will you marry me?"

She was not flustered by the question. But her answer seemed strange.

"Take off my cape," she directed.

Without saying anything, he did as he was told, unbuttoning her cape and then hanging it on the coat rack.

"Now look at me." She was wearing a sleeveless blouse that completely exposed her shoulders. She held out her stumps. In the dim light they resembled the paws of a hand puppet. Each had a scar across the end with stitch marks. They seemed hard and boney, below the muscles of her shoulders, which seemed normal.

"Can you really imagine how helpless and useless I am? There is really very little I can do for myself. I can turn the pages of a book with my toes, but little else. My mother will still have to undress me. Someday, I will be old and ugly. Can you bear to look at me then? Have me mistaken for a sideshow freak?"

He was not quite sure what to say. Even a lawyer runs out of words sometimes. He might have pointed out that her mother would not be able to look after her forever. Instead, he said. "I have considered it all, and yes I can, and I will."

He grasped her by the stumps and held her close and kissed her again, "Otherwise, I would not have asked, nor would I have lingered so long these past months."

"Let me get my breath," she gasped.

"You haven't answered my question!"

This time she kissed him with a quick dart of the head like a seal catching a fish. Then she said, "Of course I will."

There was a large wedding not long afterward. Some remarks were made about why John would marry a hopeless cripple and why had he not reached higher into the social register. The elderly couple who took care of the house promised all the help they could give. They liked the young lady although they had hardly met her.

There was another major source of gossip. Isadore and Bubbles were not there, despite an invitation. They had married quickly and quietly and were far away in Miami. It was the first time they had seen Mrs. McGinnis smile in a very long time. Dr. Sullivan had been best man.

It was the first time in several years that John had met with any of his relatives, and they were few. They found that his years in the war, and his isolation had changed him. He had developed what they considered an unnatural concern for the world's underdogs, something that would put him on the other side of political fences for the rest of their lives.

There is a bit more to the story, although this is about as much, if not more, than Dr. Sullivan was able to tell Dr. Smith about his friend John Fredrick, the famous Washington lawyer and New Deal Braintruster on labor matters.

John and Martha went to a hotel in Boston not far away. It had been a long day but they were too excited to consider sleep. Still it was the proper time to get undressed and go to bed. Her mother had changed her clothes after the wedding and sent them on their way. It had been on his mind, if not hers as well, that from now on he would have to care for her. That started with undressing her for bed.

He was inexperienced in matters concerning women. He had neither sister nor brother; no father nor mother, and later not even a grandmother. He had almost become engaged to a young lady prior to going off to the war. But she had married another and he had not regretted it.

Dr. Sullivan, or Sgt. Sullivan did not mind chasing the fair sex, and sometimes used the services of ladies of the evening when on leave. Sometimes he drank a bit too much. John Fredrick did neither, although he did not reproach his sergeant for doing so.

"Time to undress me for bed," she said.

"Where should I start?"

"Where would you like to start?"

"Perhaps I should be first."

He sat on the bed and removed his shoes, then his necktie, and then he started towards the bathroom. He came back a moment later carrying his trousers and his shirt and wearing a dressing gown.

She gave him a reproachful look as if to say, "That's unfair," but said nothing.

He hung them up. Then he turned to her and said, "I think I'll start at the top."

He took off her cape. She was wearing the same sleeveless dress she was wearing the night he proposed. He tried not to look at her stumps. She turned around and he began to unbutton her dress. He pulled it down and she stepped out of it. He hung it up. She had on a half slip underneath. He stopped and looked at the smoothness of her shoulders. He studied the latch work of her brassiere strap and unfastened it. She turned around.

He stared at her. He had once before had a fleeting glance at her bare back and bosom but then it was forbidden.

Now was different. She found it a bit alarming, as if something were wrong.

Somehow she seemed very different from the Venus de Milo he had seen years before when he had been on leave in Paris. Somehow looking at her was much more exciting.

Then each of them surprised the other. He asked, "May I touch them?"

"Of course. You will have to touch every part of me." She almost added that it wouldn't all be pleasant either.

Perhaps she had expected a timid tap upon a nipple. Instead, he seized both her breasts, one in each hand, perhaps as one would pluck grapefruit. He held them for a moment, not squeezing, but not especially gently either.

Whether it was uncomfortable, or perhaps an affront to her maidenhood, or perhaps impulsive impishness, she then surprised him. She seemed to shake herself, and her breasts seemed to come to life and squirmed out of his grasp like frisky kittens. Once free they flew about in circles and figure eights for a few moments.

"I'm not that helpless." She was laughing. His hands, his eyes, and his mouth all sprang wide open.

"What? How? Quick, do it again!"

So she did.

"If a kitten can catch a butterfly," he muttered and suddenly reached out and caught them both in mid flight.

The mammary tumult ceased. She gasped, "I yield." She was laughing too hard to say anything, although she was not very loud. Her breasts seemed to have grown larger and he wondered if it was really so or if his mind was playing tricks on him. He had not imagined how soft and warm they would be, or how large.

He pulled her closer and bending over, still holding her by the breasts, held them to his cheeks, and then kissed them as if they were the hands she no longer had. Somehow she nibbled on his ear and whispered, "There is more of me."

And of course there was.

A few days later she surprised him again. They had come from the theater and as they passed by a dark alley, a man suddenly jumped out and said, "Put up your hands!" The man held a small revolver.

There seemed to be no one else on the street. The robber had not even gotten around to demanding money when he saw that John had put up his hands and Martha had not. Even if she had put up her stumps, it would have only raised her cape as high as her ears.

"You too, lady," said the robber motioning toward her with the gun.

"Now see here," said John. The gun swung back towards him.

At that instant, her left foot shot up in an arc, striking the robber's wrist. The gun flew out of his hand across the street into the gutter. As she brought her leg down she struck him in the stomach with her foot. This doubled him over. With another step, she kicked him in the head with her right foot. All this in much less time than it takes to tell it. It was like a secretary bird striking a snake. The man fell and lay motionless on the sidewalk. She exclaimed, "Run, before he wakes up."

She could run, but poor John could only manage a fast hobble. They did not report the incident to the police, although he was troubled about not doing so.

What was worse was the item in the paper the next evening reporting the death of a man wanted by the police in the same area that they had encountered the robber. It said he had been badly beaten, but did not say it was the cause of death. He hid the paper from her. It was surely self defense. After all, a coroner's report needed to precede any cause of death report. A day later he called a police friend who told him that the man had been beaten first, and then stabbed. The robber had many enemies, it seemed.

When they returned, John set about arranging things to make it possible for his wife to open and close doors, turn on water faucets, and do many other things the rest of us take for granted.

But that is another story.

The Tooth, The Whole Tooth,
& Nothing but the Tooth

The day had come for Dr. Smith to work at Old Nasty. Fortunately, he had the company of Dr. Sullivan and Mr. Reilley. Mr. Reilley was another government inspector who would work on the kill floor with Smith and Sullivan.

Smith had heard many strange and disturbing tales about the old place and the men who worked there. Attilla Farkas, who ran the kill floor and sometimes served as the man who actually killed and bled the animals was a Hungarian who had served in the Balkan wars. His partner was nicknamed Lazar the Lender. Lazar Weissmann was a Jew from Jassy. He had employed other names in his past. But Lazar Weissmann was how he was known to the authorities.

When Attilla and Lazar argued, which was often, they would speak in Romanian, or sometimes Hungarian, which few could understand. Their English was quite adequate, however. Weissmann handled the money and sales affairs of the plant. Farkas ran the plant.

Attilla Farkas was a very large man, even larger than Dr. Sullivan. He was very strong. He had jet black hair and eyes. His skin was dark and swarthy. He seemed to have perfect teeth that gleamed white in his otherwise dark face when he smiled. Some said that he was a Gypsy.

When Lazar Weissmann smiled, the gold tooth that replaced his front upper incisor shone. When and if either of them smiled, one suspected that something was afoot or amiss.

Many of the employees had police records. Some had served time. Lazar Weissmann was reputed to lend money at high rates to desperate people and sometimes use violent and extreme means to secure repayment. Each in his own way was a man to be feared.

One of the workmen, joking with Smith, asked if he had seen the horse hanging in the number two cooler. Smith said, "No," realizing he was being kidded. He did mention it to Dr. Sullivan who said, "Lets go look. But be sure you have a good flashlight. And if you go somewhere around here, let someone you trust know where you are supposed to be, just in case you get locked in...or something," he added ominously.

Being accidentally locked into a cooler or freezer was and is a very serious matter. If you are working on the kill floor in hot weather, you won't be wearing anything like the sort of clothes you will need in a chill cooler. Much less a freezer at 0 F or less with the fans blowing.

Hanging in the cooler was an enormous carcass. It was, however, that of an enormous bull, not a horse.

"Wow, that's a big one," muttered Smith.

"What is it?" quizzed Sullivan. "With no hooves, or head."

"I can count the ribs, whether 15 in a bull or 18 in a horse."

"Yes. And horse meat is darker, and has a different smell. But once it's removed from the bones and ground it's harder to tell. But we can still tell with serum tests. Even if it's ground and mixed, we can test it in the lab to determine species."

When they got back to the government office, Smith asked, "Have you ever eaten horse meat?"

"Yes, I have. I found it tough, sort of stringy, and of slightly different flavor, but at the time I was damned glad to try it."

"Tell us about it," asked Reilley and Smith.

"Well, Lt. Fredrick and I had been making our way back to our trenches when the German attack came. We were pinned down for several hours in the barrage, and when it was over we found ourselves in a shell crater about forty feet across with a couple of Frenchmen. Also, inside the crater were a freshly killed horse and its German rider. I remember him well, lying there with a tiny but fatal hole in his forehead.

"The Frenchmen glanced up at us and saw that we were Americans. They had begun to remove some skin from the haunch of the horse, and then cut out a chunk of meat. They built a small fire and began to cook the meat. I knew that it is the custom of the French to eat horsemeat. One of

them gestured to us that we should join them if we were hungry. Lt. Frederick did not like the idea and remarked that perhaps we could start on the German after we had finished the horse.

"To our surprise, the French soldier answered, in perfect slightly accented English, 'By the time we finish the horse, the Boshe will have spoiled. Tragic, no? Here I am, a trained chef with my assistant, a splendid horse, and no proper seasonings for the meat.'

"Anyway, I tried it, even if Freddy wouldn't. I guess I really wanted to know what it was like. I have forgotten his name but the Frenchman asked me to dinner. Then he would show me what really wonderful things could be done with horsemeat...if we should we both survive the war. Too bad I never saw him again afterwards."

The first day at Old Nasty had passed without incident. To Smith, Farkas and Weissmann had seemed normal and decent men. He had said as much to Reilley.

Reilley answered, "You just wait until Farkas gets into one of his strange moods."

The next day Smith was cutting cattle heads when he felt a strange twinge of pain in his lower jaw. Perhaps it was a bad tooth. Smith had come from one of those places in Texas where the water contained just the right amount of fluoride and had never had a cavity in his life. The pain went away, but returned during the night. This pain came and went during the next week. He mentioned it to Dr. Sullivan. who suggested that at his age he just might be teething.

That is, his third molars, sometimes called wisdom teeth, were erupting. Sullivan suggested that perhaps a dental visit and x rays were in order.

So Smith took a day off to visit the dentist. He sat in the small waiting room reading for a few minutes until aroused by a pleasant voice saying, "Young man, you are next. I will need some information about you first."

Standing there was an ungainly young woman who Smith assumed was a nurse or dental technician. He was wrong.

"Good morning. I am Dr. Elsie Hershfeld."

For a moment he was not quite sure what to say, except for, "Good morning, ma'am."

She was tall, with reddish brown hair, freckles, and the air of a school teacher. She would allow no foolishness, he was sure. She seemed to be all hair, hide and bone, like an old dairy cow, a Guernsey cow to be exact. And he noted, she was the dairy type, being of massive and perhaps pendulous bosom. She had large strong hands and fingers, he noted.

"Sit down," she commanded, indicating the dentist's chair. "Now, tell me about it."

So he did. She took a series of X-rays. She poked about with dental picks, and gave his teeth a thorough cleaning while waiting for the films to develop. Apparently Dr. Sullivan had diagnosed the problem correctly for she finally announced; "You have no dental caries (cavities) that I can identify but..." she stopped for a moment before pronouncing sentence, "you have four impacted third molars that should come out."

"Will next Friday afternoon be a good time for you? One at a time will be best. Since they are impacted we may need to give you some recovery time between extractions."

"Yes," he replied, "it will give me the weekend to recover a bit."

Smith's mother had had her wisdom teeth out at the turn of the century. Then she had considered herself lucky to have had nitrous oxide anesthesia. He was not surprised at Dr. Hershfeld's findings. Other than that, neither of his parents had had much to do with dentists, nor had he, a matter of good fortune and geography.

The week passed quickly and uneventfully. He found himself back in the dentist's chair soon enough. She removed the first of the four impacted wisdom teeth. It didn't seem nearly as bad as he had imagined it would. At least it didn't while he was in the dentist's chair. The next few days weren't so pleasant, though. He had to watch what he ate and keep the empty socket clean. Avoiding infection was very important, for this was before there were antibiotics and sulfa drugs.

Another week passed and another tooth was due to be removed. Dr. Hershfeld examined the site of the first extraction and decided it looked good and began preparation

68

for the removal of the opposing tooth. It might hurt a bit more afterward, but he would be able to chew with the teeth on the other side of his mouth. Then perhaps more than a week would be allowed for his mouth to heal before the two teeth on the opposite side would be removed. So there was a visit which was merely for checking on the healing of his jaw. He asked her how and why she became a dentist.

"How did I happen to become a dentist?" she answered. "My uncle is a dentist and I work with him. Though he is away now on vacation, he will be back next month. I worked in dental technology with him first, and then later went to dental school. Perhaps I'm older than I look."

Smith remembered that the sign had said Hershfeld and Hershfeld.

"But there is another reason I got interested in dentistry. I grew up on my parents' dairy farm not far west of Chicago. One day when I was driving the tractor, the front wheel fell into a hole and I tried to take a bite out of the steering wheel. I lost my upper incisors."

She pushed with her tongue and her upper front teeth came loose. She took them out, and then put them back.

"I made these myself. I made dental appliances for a couple of years before I went to dental school."

"So I'm more like a cow than you might have imagined. I was in high school and since I was homely anyway, I got teased a bit. It was a bit hard to be named Elsie, too."

Smith looked up at her from the dental chair and said nothing. He certainly had not realized that her front teeth were false. Somehow one got used to her appearance. He certainly would not say she was ugly. He thought she had a certain beauty that was her own. When she smiled it was like a sunrise. It changed everything about her.

He asked her about her family and her early life. There wasn't much he could say while the work was going on in his mouth. All he could do was listen. He might have wanted to tell about himself but his mouth was plugged. He thought about the two remaining teeth to extract and the two follow up visits.

Then he would never see her again. He had made the rounds of the churches, but had not found any girls that he liked or that liked him. Perhaps it was his profession. He wondered if he still smelled of the packing house and the stockyard. It had not occurred to him to visit a church of a denomination other than his own. After all, there were many Baptist churches, and he could spend the greater part of a year making the rounds. He realized he was not handsome. There was a bulldog look about him, that to his credit, was more than skin deep. When he found himself in difficulties he simply hung on until he prevailed. Though while he was young he resembled a Boston terrier, neat, compact and muscular, with a strong jaw and slightly bulging eyes. As he aged, he resembled an old English bulldog. Somehow it was fitting.

During the week, he found he was humming to himself a naughty ditty he had learned as a boy. Then he had not thought there was anything naughty about it, for he was too young and innocent to know about double meanings. Later he realized that one could say very naughty things without using naughty words.

"How now, brown cow?
Get in your stall.
What thinkest thou,
with your pretty brown hair
and dark brown eyes and
winsome air.

How Now,
Brown Cow ?

70

Sometimes I wish I were the bull.
How I like to squeeze your teats,
pull and squeeze 'til the pail is full.
Toss your head and swish your tail,
but I pray don't kick that pail."

So it was that he made his first visit to a Lutheran church. Not the one that Elsie attended, but another where English rather than German was spoken and where if he made a mistake, she would not learn of it, at least not immediately. A practice session, he thought of it. He was surprised how friendly the people were. He was a bit concerned that perhaps he was not playing fair with them, since he had an ulterior motive.

It was not that he was aware of having any romantic notions. He knew she was a bit older than he, almost a year, in fact, and that by rights, she was professionally his superior. But she was nice to be around and talk to. And that was an important reason to want to see her. For he had almost no one to talk to about anything but work, and after nearly two years, it was almost like Robinson Crusoe seeing the footprint in the sand.

It must have been the week before he was going to have his last wisdom tooth removed that he went to the same church as she had said she attended. Alas, she was not there. It turned out that she had gone to the station to meet her aunt and uncle returning from a long vacation, his first in several years.

He did not tell her about his church going activities, at least not then. He had not quite worked up the courage to ask her for a date, nor did it seem quite proper so long as he was her patient. Besides something else was on his mind.

His mind had been away from his work at the plant during the time he had been having his teeth extracted. His work at Old Nasty had gone along without any major problems. If there had been any problems, Dr.Sullivan and

71

Mr. Reilley had taken care of them without Smith being aware of it. But now Smith seemed to sense that something about Mr. Farkas was different, or at least changing. Remembering back, a few weeks before, Mr. Farkas had seemed shy and retiring. Now he seemed outgoing and expansive, at times a bit overbearing. As for Mr. Weissmann, he was the same sly and evasive man he had always seemed. Smith decided he was letting his imagination run loose and put it out of his mind.

One day when they had finished inspecting pigs, Dr. Sullivan had remarked about the pig's pancreas being used for medical research and for the then new commercial production of insulin to treat diabetes.

"I never said anything to you about it. I guess you know my wife died from diabetes. At least I think she did. I never really was certain. But I think about it often. We were married when I was just out of school, high school, that is. That was more than twenty years ago. For a year and a half we were very happy, then one day she was going to have a baby. I was twenty. She was eighteen. I was working in a packing house then as a regular employee. I knew nothing of medicine. I had never been sick a day and had little interest in such matters.

"However one day I came home and she was deathly ill. I sent for the doctor. She was taken to the hospital. It seemed that she had borderline diabetes, and pregnancy was too much for her. She died a few days later.

Had she lived I might have made a good packing house foreman, or perhaps a plant manager. I doubt I would ever tried to go to veterinary college. I resolved never to love anyone or anything very much again. I guess it helped me during the war...I already knew the shortness and precariousness of life.

"So for a very short time we were very happy together. Then she was gone."

He sighed and turned away as though he were putting something away.

Smith really did not know what to say under the circumstances. It reminded him of the story of Jurgis in "the Jungle", a novel he had read. Jurgis, the Lithuanian immigrant who had lost his wife in childbirth, had been a packing house worker.

The weekend brought a happier time. On going back to the Lutheran church this time, there she was. Somehow, she looked beautiful. Perhaps it was the change of clothes, perhaps it was the change of place, perhaps it was a change in him.

So he got to meet her uncle and aunt. Then there was an invitation to supper. Something had started.

She liked him, but did not take him seriously. He seemed too young. Perhaps he was, but he could be handy to have around for at the moment there wasn't anyone else. Conversation around the dinner table that evening concerned life on the farm, veterinary medicine, and their respective families.

Smith liked her aunt and uncle immediately. It set him to wondering if he should have been a dentist.

It seemed very confining though; being in the same office all the time, like small animal medicine. Rarely could it be as grim and grimy as being a large animal veterinarian - or a meat inspector.

One evening, a few days later, a Catholic priest called on Dr. Sullivan at his boarding house. When they had gotten seated in a private room he introduced himself at greater length, "I'm Father William O'Roark. I've come to ask about someone you know and work with. I'm concerned because someone has confessed disturbing things to me, as well as asking my help."

Dr. Sullivan gave him a disturbed look for he did not want to know anything that might compromise a man's

73

confession. Father. O'Roark must have read the look correctly, for he said, "No, the person I need to know about hasn't been to confession in a great many years. It was a member of his family that asked me to look into the matter."

"I see," Sullivan said. He wondered if he were the person referred to for he had not been to church or confession for a very long time. Not since a friend's funeral some years before had he been. Besides, if he had any family, he was unaware of it.

"Do you know Mr. Attilla Farkas?"

"Not well. He is a hard man to know. Last spring I thought I knew and understood him pretty well but now I'm not so sure."

"Someone in his family is very concerned about him."

"You mean, to get him into church?"

"Yes, but to keep him out of trouble, as well."

Father O'Roark meant jail, but did not say so.

"I'm really not at liberty to talk about the people I work with. I have somewhat the same handicap you have Father."

"I see. I can tell you his wife has had to leave him, because he has become increasingly violent in the past few weeks. It appears this is not the first time it's happened either. She will return when she thinks he has calmed down."

He paused a moment as if something else was on his mind, "Somehow you seem familiar to me. Could we have met before?"

"Strange, I had the same feeling," Sullivan answered. "Were you in the War?"

"Yes, I was a chaplain in France, during 1918 and in the hospitals after the Armistice."

Conversation began in earnest. They could trace their movements during the war but could not quite locate a common meeting ground. They both sat silently for a moment. Then the answer came.

"Where did you grow up?" asked Father O'Roark.

"In an orphanage just outside of Boston."

A light seemed to shine in O'Roark's face as he exclaimed, "Why, you're Paddy Sullivan! After all these years! So this is what has become of you. The last time I heard of you was when you married Mary Feeny. How is she?"

"She's been dead years, Bill. It sometimes seems like yesterday - even after twenty six years."

"I'm sorry. Can you tell me about it?"

He began to tell him the same story he had told Wilber Smith the week before but, in greater detail. It was a long soliloquy which he ended by saying, "I shed all my tears then. I swore never to love anything or anyone again so much as I did her. It left a great numb dead place in my heart. I went through the war, saw terrible things, friends die in agony, towns destroyed, and it never really reached me. I know their suffering must have matched mine but I did not feel it. Not that I wanted to.

"At first, I wanted to drink myself to oblivion. I lost my job. Suddenly, I came to my senses and decided it was stupid to punish myself, for I had not done anything wrong. I went back to work with the aim of beating back this cruel world."

"Did you ask God's help?"

"No. I just asked that he leave me alone and not hinder me. Perhaps I prayed in the trenches for myself, but it was just fear."

Sullivan hesitated and then said, "I haven't been to confession since my wife died. I really haven't forgiven God, you see. It would be strange to ask Him for forgiveness, for my small sins. Is He sorry for what He did to us?"

75

Sullivan stopped and thought a moment. He could not believe he was saying all this, for it was much more than he had ever said to anyone about himself and his feelings. He continued, "So I suppose you have heard my confession, or part of it. I have gotten drunk on occasion, had carnal knowledge of various women when the urge and opportunity coincided, even been in a fight or two.

"So now, before you can tell me what penance I should do, you might suggest to the Almighty, what penance He should do."

There it was; the entire blasphemy he had blurted out in a quiet and restrained but very angry voice.

They were both very quiet. There wasn't any quick easy answer.

"Have you asked God for an explanation for your suffering and Mary's?" asked Father O'Roark.

"No, not really, I suppose. I know that the world is full of undeserved suffering. Even the pigs we kill probably deserve some better fate, though I suspect they die less painfully than we often do."

He stopped, and seemed to pull himself together, to collect his wits, and redirect his thoughts to happier things. "Tell me about yourself, Bill. How do you get along without the fair sex? Or do you?"

It was an unfair question. Had he thought about it he might never have said it. Father O'Roark gave him a searching look and then seemed to come to a decision.

He replied, "Actually, it isn't the fair sex that attracts me, Paddy. It's my own. But I haven't yielded to temptation since Sister Periwinkle caught us in the hay loft that afternoon so long ago."

Sullivan had forgotten the incident, or repressed it well, but now he remembered. He felt a strange mixture of admiration and loathing one sometimes feels for the cruelly afflicted; loathing for the affliction and admiration for the victim struggling against it.

"I have prayed to be relieved of it but it hasn't gone away yet. I keep thinking, perhaps tomorrow..." Father O'Roark's voice faded away in deep thought.

The clock struck nine.

"I must be going. It's late. Pray for me Paddy, if not for yourself. I will pray for you."

Sullivan hugged him for a moment, they parted and he was gone.

The rest of the week passed and Saturday night brought him a terrible dream. He was seated at the kitchen table. He was half drunk. The whiskey bottle in front of him was half empty. The kitchen was dirty. He thought a rat ran across the floor, but he couldn't see well enough to be sure.

He looked up. He saw his wife, Mary, but she looked very different from the way he remembered her. Her hair was gray, stringy and knotted behind her head. She was much older and fatter. Her face was red and wrinkled. She was ugly.

She was angry with him. There were many reasons. He had been drinking again. She had a black eye he had given her. She was crying. Evidently, this was something really worse than the poverty that afflicted them.

She sat down, folded her arms and buried her face in her arms folded upon the bare table. There was the newspaper on the table. He should have wanted to hold her again and kiss her, but somehow he had lost all feeling for her. It seemed strange as well as horrible. He managed to read the headline on the paper in front of him.

It said, "Sullivan Boys Electrocuted."

He knew he had read it before. He half remembered how the three boys had gone wrong, how he had failed as a father, how they had committed first crimes against society; first robbery, and finally murder.

She was saying she wished they had never been born, that she would rather have died than give birth to them. He seemed to agree.

He awoke in a cold sweat, and could not go back to sleep. It all seemed so real and so horrible. He got up and lit a cigarette and sat in his chair. There was a certain horrid logic to it. Perhaps it was an answer of sorts.

He went back to bed and slept fitfully. Now he dreamed he was sitting in a doorway with another old man. They were arguing about whose turn it was to drink, for the wine bottle he clutched was nearly empty.

He looked down. The was a piece of newspaper on the pavement. The headline said, "Hitler invades." He could not make out the rest. A police car drove up. There was an unfamiliar streamlined look about it. The policeman got out and started to walk towards them.

He awoke again. For a moment he wondered who Hitler was. Then he remembered that Hitler was the new German Chancellor. A few moments passed and then he heard church bells ringing. He thought, "It's Sunday, I will go to church today." It seemed a novel idea, but one whose time had come.

Somehow the world looked different to Dr. Sullivan. He spent much time thinking about his past and looking inward at his own thoughts. He seemed preoccupied in a way that he had not before. He had been a man who lived in the immediate present and seldom thought of past or future. Now his life would soon change, partly because of this change, and partly because of something which was about to happen.

Perhaps it began one day when there had been a long and difficult day working on the kill floor. Smith and Sullivan were both very tired when it was all over. The animals had been bad. That is, several of the cattle had tuberculosis and they had to do detailed examinations on four of them, as well as the paperwork. There had been machinery breakdowns as well. The refrigeration equipment had leaked ammonia and the men had been driven off of the kill floor while other men with gas masks had gone in to repair and seal the leaks. Tempers were short.

It happened that Smith was leaving a few minutes before Sullivan. So Smith witnessed an incident which a few weeks before he could have scarcely imagined.

One of the men had accused Lazar Weissmann of cheating him on his pay for the day. He pulled out a knife, and began to threaten those around him. Perhaps he had been drinking.

Farkas came upon the scene, and distracted the man, and then gave him a thorough beating, finally dragging him out the front gate, bruised and unconscious. Farkas had, however, stopped and asked Weissmann what the man thought he was due. He then took money from his own pocket and shoved it into the other man's pocket. Perhaps it was more than the man had claimed. Soon the man revived, got to his feet, and stumbled off towards the streetcar stop.

While Smith happened to see the incident he had no cause to involve himself. He had no idea as to who was at fault. Farkas and Weissmann exchanged angry words but since it was not in English, Smith had no idea what was said.

Fall came and Elsie invited Wilber Smith to go on a hayride at her church. They found themselves in the back of the wagon holding hands. Wilber began to hum a tune, "How now, brown Cow?"

To his surprise she began to sing the words. He joined in. Somehow though, when they got partway into it she got very red in the face and the words seemed to die away. They looked at each other, both blushed very red, and then began to laugh. It took them a while to settle down.

79

They could give no good explanation as to why they were laughing...at least not at a church picnic and hayride.

Late one morning, Mr. Reilley and Dr. Smith were examining a suspect cow, one of the cripples that Lazar Weissmann bought on consignment. The "I'll pay you if the vet passes it." kind.

A farmer walked into the scale office and began to tell Lazar Weissmann a strange tale, "I've got some pigs I need to do something with."

"That's Okay by me. The meat's treff but the money is Kosher." said Lazar. "Send 'em in here. We'll give you an honest price on 'em."

"Well, you see, these pigs are special," said the farmer.

"Oh. You mean there's something wrong with them?" Lazar looked around and noticed that Smitty and Reilley were too far away to hear. "You mean they ain't really yours...or maybe they are under a quarantine, or maybe sick?" he said as he stroked his stubble covered chin, a faint grin crossing his face.

At that moment Smith and Reilley turned and came towards the scale office.

"That cow won't make it, got a temperature of 106.5 along with mastitis and pneumonia," said Smith. "She's condemned. I'm sorry."

"Okay, I'll get Farkas to kill her and haul her directly to the tank."

A generation later, there would have been sulfas and antibiotics available that could have been used to treat and possibly save a cow like this one. But this was 1934.

The farmer said, "Seeing you are the government inspectors here I'll tell you about the pigs, too."

"Okay." said Reilley.

"You see, about a week ago poor old Uncle Henry was out in the hog pen with his hogs and he had a heart attack, we think. Would have been 90 his next birthday. He was a widower living alone, and nobody knew anything was wrong right away. When Carrie, my wife, went over to see him, she couldn't find him. Finally, our hired man found him, or part of him in the hog pen. He was pretty well et up by then. Couldn't hardly find enough to properly bury.

"Everything was left to us being he had no children. So the hogs are ours.

"But I don't know just what is the right thing to do with those hogs. I thought of shooting the lot of them, burying them on the place along with Uncle Henry, so he would be all together, you might say. Is there any law about it?"

"Not that I know of," said Smith. "You know of anything?" He glanced at Reilley

"Nope," said Reilley.

"Obviously, a man wants to do the right thing under such tragic circumstances," stated Mr. Weissmann. "Such matters however, require a proper point of view."

He gazed upwards as though receiving divine guidance, holding his hands out palms up. "Suppose your poor old Uncle Henry, God rest his soul, had passed away and you had buried him under an apple tree, would it bother you to eat the apples from that tree?"

The farmer's face brightened as though a light had begun to shine within. A few moments passed and he spoke. "I'll bring them in tomorrow!"

A few days later Elsie gave Wilber Smith a strange and interesting trinket, something that only a dentist might really appreciate.

"What is it?" asked Wilber.

"A tooth, the whole tooth and nothing but the tooth," she quipped.

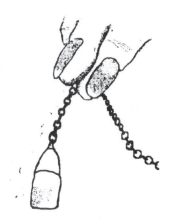

And so it was. A tooth, made of porcelain and gold, looking ever so natural and real, an incisor with its single root and covered with gold over most of its enamel surface. But it had never been in anyone's mouth. It dangled from a gold chain and had been made for an ornament, "This is an example of the sort of work I did before I became a dentist. Too bad I don't know how to make teeth we can put back in and make them stay. But this one looks real enough, doesn't it?"

Smith was so surprised at the gift that he had a problem presenting his gift to her. He had composed a little speech but he had become confused. So he merely said, "Funny thing. I have a gift for you. But mine has conditions attached."

"Oh," she said.

With that he opened a tiny box, and took out a gold ring with a single diamond. He wondered out loud, "Will it fit?"

"We will make it fit," she said in a determined voice. She started to put it on but it was too small. "Now what was the condition?"

"Perhaps I should ask your father first."

"My uncle will do."

The answer to the question was, of course, "Yes". From both. As if you couldn't guess what the question was.

Smith had saved about three weeks of vacation time most of which he would use for the wedding and honeymoon. Just before he left the plant, on Friday afternoon before leaving on his wedding trip, he reached into his pocket to show Dr. Sullivan and Mr. Reilley the porcelain tooth on its gold chain that Elsie had made and given him.

Alas, there was no tooth, only the chain. So he said nothing to anyone about it, except Elsie. He considered it a bad omen, but he felt he had to tell Elsie. Except for this last day on the job, he had not kept it with him. He believed it was too precious to carry about as a good luck piece. She forgave him and passed it off quickly saying she could easily make him another as soon as time permitted.

The wedding would take place on Monday at her parent's farm near Chicago. His parents and his brother were there and there was a great family gathering with lots of food and lots of talk. Wilber and Elsie went off to Chicago for the honeymoon. Somehow, it seemed unfair that none of his friends from work were able to be invited. Not even the nice lady at the boarding house whom he was fond of and who was delighted that he had finally found a nice wife - even though he would no longer be staying with her.

Losing the tooth was a bad omen, but not for Wilber or Elsie. But it was less than a week end later that the missing tooth began its work.

It was Saturday night. Wilber Smith was on the train to Chicago. Elsie Hershfeld was at the Hershfeld farm getting ready for the wedding.

At the same time, back at Old Nasty, there were just the four men gathered in the upper room. "The other office". It was a sort of secret place (secret to the public), yet its existence was well known to most of the plant employees. Farkas, Bronsky, Weissmann, and Mann were playing poker and other games of chance. If poker became too slow and tedious they might simply cut cards, with high card taking all.

Mann was not exactly part of the inside staff. Neither for that matter was Bronsky. A major part of their duties involved other activities outside the plant. Collecting bad debts was one of them. Revolvers in shoulder holsters were often part of their attire. Just now they hung from the hat rack on the wall. It was a place most of the employees regarded with fear. One taken there might come out in much worse condition than he went in. There was only a small skylight, which dirt nearly hid, and the room was relatively soundproof.

The other three men dwarfed Weissmann. Weissmann, however, knew that he had more brains than the other three put together. Often he let them win money he could easily win back from them. Farkas, however, was again becoming a problem. He was becoming willful, headstrong, and foolish again. He might do something rash and foolish that would bring the wrath of the authorities down on them. He would have to be dealt a lesson. It was not to work out that way, though. Farkas was a wild card.

Sullivan heard about these games. He would have really enjoyed playing a game with Lazar Weissmann; but his common sense absolutely forbade it. He warned his inspectors against ever playing games for money with anyone connected with a plant. It was just too easy for something to go wrong. If you lost, they had something on you, a debt they could use to manipulate you. If you won it was a sort of bribe. Certainly it was much wiser not to get too close to people on whom you had to lay the law down.

This night, Weissmann only pretended to drink. Farkas and the others drank heavily. Bronsky and Mann won a little, lost a little, but managed to gain a bit at Farkas' expense.

Farkas seemed to lose steadily. Yet he kept increasing his bets, believing he could make up his losses this way when his luck turned.

Soon Bronsky and Mann dropped out. They realized they were winning more than it was safe to win and that something was going on, something unusually sinister.

Now, Lazar Weissmann usually played his victims like a skillful fly caster plays a trout. This night something was different. Perhaps he really wanted to teach Farkas a lesson not soon forgotten. Perhaps the quarrel over the man's pay a few days before was still burning within him. Farkas did not have the kind of money to keep up with Weissmann, even if he had been as skillful at cards, which he was not. Neither was he sober.

Weissmann leaned back and remarked, "That's about twenty five grand you're down. Shall we call it a night?"

"No," sneered Farkas, "lets cut cards for it. Double or nothing."

"You sure you want to do that? That means your half the plant belongs to me if you lose." He sighed. He was riding the tiger and dare not dismount. He suddenly wished he were somewhere else. He just would have to try to lose this one, or should he? There were times (and this was one of them) that would have done anything to be rid of Farkas, however useful he was at other times. If Farkas won now, he would be insufferable.

"Okay, you shuffle, Farkas."

Drunkenly, Farkas shuffled the cards.

"Go ahead and take the first cut."

He did and drew the ace of hearts. A triumphal sneer crossed his face. Clearly his luck had changed.

He leaned back and handed the deck to Weissmann. There could be only one higher card in the deck, odds of 50 to one in his favor.

Bronsky and Mann, having collected their more modest winnings, stopped in the process of putting on their coats over their shoulder holsters and stared.

Cautiously, Weissmann cut the cards. There was the ace of spades. Perhaps he had just time enough to remember that Carmen in the opera had drawn the ace of spades predicting her death. For in that instant, in one motion, Attilla Farkas picked up the heavy oak table and hurled it over onto Weissmann, smashing him into the large iron safe behind him, shouting in maniacal fury, "You cheated!"

YOU CHEATED!

Perhaps it was true, for Lazar Weissmann had cheated often, and in many ways. Perhaps he had really cheated the man of his just pay when Farkas had possibly saved someone from being stabbed.

For a moment Bronsky and Mann stood transfixed.

Farkas immediately pulled the table from Weissmann in regret at having injured his friend in a fit of rage. He cried out, "Help me get him up!"

Their efforts were useless. Lazar Weissmann was dead. The angle of his head showed that his neck was obviously broken, and he had a massive skull fracture in the occipital region. Either one alone would have been fatal.

Farkas pleaded with him not to die. Slowly, Farkas realized that Weissmann was dead. Then Farkas stood up and turned away towards the wall. He covered his face with his hands and began to weep. He had forgotten that a few moments before the same man had ruined him financially. Such was the mercurial nature of his illness.

If it occurred to either of the other two men that, with his back to them, Farkas was an easy target for the revolvers they carried, they gave no indication of it.

Mann finally broke the silence, "Boss, it was an accident. You didn't mean it."

Such sympathy amongst men who routinely dealt out physical harm seems ludicrous, but it was genuine. No one considered calling the police. They all sat down at the table, now set back aright, and attempted to make plans, each having poured himself a healthy drink from the bottle which had survived the fall from the table.

Lazar the Lender lay dead on the floor beside the table.

They finally decided that Lazar Weissmann should be put away as whispered legend said had befallen some others...in the rendering facility in the basement. So far as they knew Weissmann had few friends and no relatives in America. If he were said to have gone back to Jassy, who would question it?

They divided the money that had been scattered on the floor, several thousand dollars.

Next, they cleaned up the room as best they could. Then they carried their victim to the tanking facility. Then they removed most of his clothes and stuffed them into the coal furnace under the boiler. They checked and emptied the pockets first, however.

Farkas kissed the corpse's bloody cheek, wiped away his own tears and took a drink from the bottle. Then he turned the body face down and began the job of dissection, using only a very sharp knife.

Expertly, he separated the head from the neck, cut the legs apart at the knees, removed the arms at the shoulders, and the thighs at the hips. Meanwhile the other men had started a fire in the furnace under the boiler. Not long after midnight the funeral was complete. Mann and Bronsky left. Whether they slept well one doubts, but Farkas stayed to make sure the hammer mill on the other side completed its job so there would no revealing bones.

The ignominy of a Jew ending with his own hogs brought Farkas to hysterical laughter. Then, a few moments later, he was again weeping at his friend's fate. But, he reasoned, what else could he do? There was a strange odor about the place for men and apes are of a different flesh than hogs or cattle. It was a horrid familiar smell he remembered from the Balkan wars.

When Monday came, Attilla Farkas told everyone that Lazar Weissmann had gone back to the old country, leaving Mann, Bronsky, and Farkas in charge of the plant.

They would have to forge papers to that effect. Meanwhile a typewritten letter with a forged signature would have to do. It meant closing the loan shark business for a while, perhaps a few debts would have to be forgiven. Who would question what had happened to Weissmann?

Mr. Reilley, Dr. Sullivan, and Dr. O'Neal (a new trainee veterinarian who was covering for Smith) checked the plant for sanitation. It seemed unusually clean, especially the tank house. Farkas had done an unusually good job of it. Alone, it seemed. Again, he explained that Weissmann had gone back to the old country. It might be so or not, for Weissmann had been known to pretend leaving and lay low before when the heat was on.

Farkas seemed unusually tired, but now he had more physical help from Bronsky and Mann. What they couldn't do was help him with the sales and money end of the plant. It would take only a few days for this to bring real trouble.

O'Neal remarked that there was a strange odor to the tank house that morning. Reilley concurred but couldn't place it. Perhaps they both suppressed it. Maybe they had noticed it there before and forgotten. There had always been jokes about putting someone in the tank, but no one had ever heard of it ever really happening.

After operations were over for the day and just before the paperwork began in earnest, Reilley took O'Neal back to the tank house to explain its machinery and operation. From a doorway, Bronsky watched unnoticed by the two inspectors. He saw one of the two men suddenly lean over and pick up something tiny, something with a golden glint. At any other time he would have assumed it was a fired shell casing from the captive bolt pistol. Now, however; he feared it was something else; something that could perhaps provide lethal evidence of foul play.

The two inspectors went back to their office. Bronsky went to find Mann. He wasn't sure he wanted to tell Farkas. In his present state of mind there was no telling what Farkas might do. Bronsky and Mann argued for a minute and decided that they would have to tell Farkas. After all, he did the killing, even if it had been an accident.

Dr. O'Neal had found a gold tooth. On Sullivan's advice he left for the police station with it. Reilley and Sullivan simply sat at their desks and filled out the kill and processing reports, as if nothing else mattered. Soon Farkas,

Bronsky and Mann were at the door. They had not made up their minds just what to do. Maybe there wasn't any tooth, just a shell casing. Even if there were a tooth, who could be sure it came from Weissmann? Even if it did, it was only a tooth.

Reilley said, "Come in, we'll have your copies ready in a few minutes. Have a seat if you can find one."

"Where is the other inspector?" asked Bronsky.

"O'Neal? We sent the kid home, gave him a break. Let him learn all this a little at a time. We'll start him on paper work tomorrow. He isn't even settled in here yet. I suggested he look up the boarding house where Smith lived before he got married."

There really weren't very many places to sit, just four chairs and couple of boxes. Farkas sat down on a box and leaned his back against a wall. He was very tired. Perhaps he was now coming down from the manic state.

Bronsky finally asked, "What did O'Neal find in the tank house floor? I thought we did a pretty good clean up."

Farkas glared at Bronsky since, he, Farkas, had done all the work.

"Oh, this," Sullivan tossed him an empty shell casing from the captive bolt gun. It was true, but it wasn't the whole truth for the casing had been found in the morning. After all, the sanitation inspection had taken place early in the morning before the kill. O'Neal found the golden object in the afternoon after operations. Still, the sanitation inspection had not covered the tank room very thoroughly.

"Well, here is the kill form. Four condemnations antemortem, and three during the kill. Ought to make Mr. Weissmann happy. Wherever he is."

It must have been the wrong thing to say, for a wild look came into Farkas' eyes. He began to mutter. "They know; he knows."

Bronsky and Mann both stared at him. He was going to reveal their crime, so far so well hidden. What to do, get him out of there? How, without a scene, perhaps even a fight, could they get him away before he said too much. They had not brought their revolvers and had only their knives, as dangerous as those were.

Neither Reilley nor Sullivan seemed to notice the sudden chill nor the muttering of Farkas as they continued to do more paperwork.

Sullivan remarked, "Now that Mr. Weissmann has left, who will be in the front office? Should we look to Mr. Farkas for answers?"

"We will get someone else who can keep books," said Mr. Mann It was true that they knew several people who could keep the kind of books they would require. Their problem was how to remain in control if they did. They both knew more money was stolen at pen point than gun point. The pen is mightier than the sword.

"Lets go, Mr. Farkas, I guess it's been a long day. I got all the papers we need."

Farkas lurched to his feet, but rather than wanting to leave, he pointed his finger at Dr. Sullivan, "He knows. He knows what happened to Lazar, what we did." The wild look had again filled his face. The other four men were staring at him in disbelief.

"Are you all right, Mr. Farkas?" asked Dr. Sullivan.

"He'll be okay, Doc," answered Bronsky. "He's been working too hard recently."

"You'll never tell. I'll kill you if you tell," muttered Farkas, taking a step in the direction of Dr. Sullivan.

"Tell what?" asked Sullivan.

"How we killed Lazar Weissmann. How we killed him, and destroyed the body."

Mann snapped, "What do you mean we?" Instantly, he realized it was a revealing remark he wished he had not said.

Dr. Sullivan said, "We aren't your lawyer, nor the police. If you have something on your mind, tell them. We are just government meat inspectors."

"We'll make it worth your while to forget all of this, however..." Bronsky's voice trailed off.

"You can't do that," replied Sullivan. "It says here in the Regulations, that inspectors are not to be bribed, threatened, or intimidated."

At that moment, the door behind them opened and another man came in quietly followed by several uniformed policemen.

Farkas, Bronsky, and Mann looked around in consternation.

"Gentlemen," said the detective, "we would like you to come down to headquarters and answer a few questions. Perhaps we can clear this up."

"Do you want all of us?" asked Mr. Reilley .

"Yes, if you can come down a little later."

"Thanks, we would like to get cleaned up after today's work."

A moment later Reilley and Sullivan were again alone in the office.

Reilley remarked, "I didn't know there was anything spelled out about bribery, or especially threats or intimidation in the Regulations."

"I'm very glad he didn't ask me to show him where it said that in the Regulations. Weissmann would have made me show him. I just hope what we heard isn't so."

And it was some years later before these were spelled out in detail in the Regulations.

"But in a way, it's in the Regulations just now." Sullivan reached for the briefcase holding a small paperback book. He opened it. There lay the Colt .45 automatic he had carried during the Great War, cocked and loaded.

"I think it is a regulation we could have enforced, at the time." He pressed the magazine latch, took out the  magazine, pulled back the slide, catching the heavy copper and brass cartridge as it fell harmlessly from the chamber. He looked at it for a long moment, glad that it was still unfired, and then pressed it back into the top position in the magazine. Then he pressed the magazine back into the big automatic, and dropped the hammer onto halfcock. He looked at it a long moment and then put it back into the briefcase.

Two weeks later, Smith came back to work from his honeymoon only find the plant closed. Elsie was visited in her office by a detective who showed her a strange trinket. It was the gold tooth. She turned it over and examined it with a magnifying glass. Inscribed in the gold were the initials E. H. She could see the fractured weld where it had broken free from the chain. The detective had not told her where it was supposed by the police to have come from. He was even more surprised than Elsie.

"Incredible, I never thought we would see it again. My husband will be delighted."

So the only evidence was Farkas' confession and corroborating statements by Bronsky and Mann. The court believed them and let them off with short sentences. Farkas went to join Dave Lightfoot, (who had driven cattle for him some years before at old Nasty and had let the pigs loose to get a last drink at Paddy's) in the State mental hospital, where he remained for a great many years.

This ended the close association of Dr. Sullivan with Dr. Smith. Dr. Sullivan was moved to Washington, shortly thereafter. He advanced rapidly as he was found to have knowledge and ability badly needed during and after the war. By the time he retired he could really say that he had been "someone important."

Dr. Smith and his new wife were sent to the Chicago area. So that was the story of the tooth, the whole tooth, and nothing but the tooth.

Chicago, Chicago, I'll Show You Around

Wilber and Elsie Smith moved to Chicago. They found a place to live and he went to work at Establishment 38, (which was its official government title), Swift and Co., where he was first assigned. Eventually he was sent to other meat packing plants in the area.

The plant was larger than any other he had yet seen. Far larger than the Swift plant in Fort Worth, it was almost a city in itself. Coal was brought in by the trainload and burned for the boilers and great reciprocating steam engines that chuffed away in the basement. Long years before he had become an inspector belts had run all over the plant, turning great shafts, and producing a sound like great distant waterfalls. About the time of the first world war small electric motors began to replace the belts and pulleys.

When the sun came up early the great kill floor suggested the nave of a great cathedral. It seemed large enough to house a dirigible.

Trainloads of hogs and cattle arrived in the pens. Trainloads of meat and meat products left for destinations all over the world. Some of the meat had been made into the great variety of products Swift was known for. But a lot of meat was shipped as hanging carcasses, in halves and quarters in special refrigerated railroad cars to be made into meat products at other smaller plants or simply sold as is.

This was also a place where he might remain anonymous for his own safety after the affair in Kansas City.

Here, Wilber Smith would spend most of the war years that would come. In time, Elsie produced a son. They named him Timothy.

Eventually they were able to buy a home that suited them farther from the stockyards. When the war came, Timothy was sent to live with his grandparents on the farm.

Fortunately it wasn't very far, so weekends often brought them all together.

It had been a long hard day up to this time. What happened next disturbed Dr. Smith even more. One of the pig carcasses had been rolled out by one of his inspectors. It looked as though it hadn't quite bled out right. There was still a lot of blood left in the carcass that should have drained out through the gash in its throat. Both jugular veins and carotid arteries should have been severed. It did have a proper stick wound. Well, it seemed to, but it hadn't bled out very well. The liver had a muddy look, a look that indicated perhaps fever in the live hog. The kidneys also had retained too much blood. Reluctantly, he condemned it. Even in war time it didn't make sense to take chances.

It was a little later that someone from the kill floor whispered to him that they had killed a dead hog or two. That is, a hog already dead from some other cause had been shackled and stuck. That, of course, was absolutely forbidden. Hogs, and all other food animals, must be killed by proper slaughter methods; not cut up after having died from some other cause. Only exsanguination (death from loss of blood) was allowed.

He managed to slip out to the pens. There he saw a hog in the suspect pen that seemed to be stiff. He went over to it. It was stone cold dead. The limbs were stiff. The fat seemed of the consistency of wax. He had the antemortem inspector put on the condemned tag and get the carcass moved out to the tank house.

As he walked back to the kill floor, he saw one of the hogs in the lane get run over by the other hogs and trampled. He got closer to see what would happen next. For a while, nothing happened. Then one of the plant men came and tried to move the hog. He couldn't. Failing, he went off to get help.

Unobserved, Smith examined the hog. It was dead. He made sure feeling for any pulse. He then took out his knife and made a few small cuts where they would not be noticed. No blood flowed from these cuts. Then he withdrew to a point from which he was able to observe what would happen next.

Soon, the man came back with two companions. They picked up the hog, placed it in a wheelbarrow and headed for the shackling pen. Smith watched it disappear into the building. He went inside, and waited at the other side of the dehairing machinery. In due course, the twice dead hog appeared. It was a bit redder than it should have been. The small marks he had placed on the hog made him certain that it was the same hog that had died outside. He called over the floor inspector, explained what was going on and tagged the hog. The viscera were bloody. The internal organs were engorged with blood.

He called the kill floor foreman and told him what had happened.

If the foreman had been in on the deception, he certainly did not act like it. He stormed out of the kill floor, with Dr. Smith just behind, who had delayed just long enough to slash, and brand the carcass U. S. CONDEMNED.

When the foreman finally caught up with the three offenders, the oldest one stammered, "But he was only just killed. I mean he was only slightly dead. After all, this is wartime and we need all the meat we can get."

The war had begun in Europe, but the sneak attack on Pearl Harbor was still nearly a year in the future.

"No! Damn it! No!" roared the foreman. "Do you want our plant to lose its inspection? I ought to have you guys fired. Who ever gave you the idea to kill a dead hog? If Dr. Smith wasn't a damn fine guy he would pull our inspection over this. Don't ever do that again!"

Smith accepted that it was an honest dumb mistake. After all, this wasn't Old Nasty. It was a decent plant. He could understand how someone might think that a hog killed in an accident might be all right to eat. After all, with the war on it was hard to get good help.

Not long thereafter, the circuit supervisor happened to show up in the office and Smith told him about the incident.

The supervisor asked, "Well, Dr. Smith, how did you know the hog was dead, really dead?"

"A really stupid question," thought Smith. "This guy belongs in Washington!" But he said, "The carcass was stiff, cooler than normal to the touch, had no reflexes and didn't even bleed when I cut it."

"If you did all that to be sure the hog was dead, how could you expect three simple laboring fellows to know the hog was dead? And why didn't you simply get the inspector from antemortem to tag and dispose of the dead hog?"

"I wanted to see what would happen in the absence of any inspectors. I had a tip-off earlier." Smith was just a bit disturbed at the line of questions. "You don't mean I should have ignored this and just let it go on?"

"Not at all. I just want you to have some idea of what kind of questions we get asked in the main office when the upper plant management asks us questions. No, you are right, dead right, you might say," answered the older man with a chuckle. "If this is more than a simple dumb mistake, we can be sure to hear about it from some of the plant people and I want to be ready for them."

Smith forgot the incident for a day or so, but he decided to keep an eye out for such tricks. When the weekend came he was able to get an extra day off and went out to the farm with his wife to see her parents and their son Timothy.

Timothy must have been about five at the time. Although theirs was a dairy farm, there were a few pigs that they kept for their own use. These were generally slaughtered in the fall. Timothy had to be kept away when pigs were killed for fear he would be upset by it.

One day, Timothy watched his grandmother kill a chicken. She hadn't meant for him to watch. She took her trusty hatchet and without any ceremony, chopped off the hen's head on the chopping block. Timothy watched with fascination as the chicken flopped about as if hunting for its missing head. He began to laugh. Something about it struck the boy as funny. A chicken was so dumb it didn't know when it was dead.

So they decided to let him see the hogs killed. Grandfather took out his old twenty-two rifle and shot each hog just a little above a point where two lines running from the pig's ears to its eyes would cross. Instantly, the pig sat back, fell over on its side, making not a sound. Its feet moved for a few seconds. Then it lay still. Another man came up and cut the pig's throat. There was a great gush of blood. Then several of the men would pick up the pig and put it into a huge tub of boiling water. The women scraped and scraped until all of the hair was removed. They killed four hogs that morning. It seemed to take all morning and late into the afternoon to get everything done.

"Say, Wilber," said Grandfather Hershfeld to Dr. Smith, "You can inspect this hog for us."

"Sure," said another man, who was helping.

"Well, yes and no," answered Dr. Smith. "I can't mark them U. S. Inspected and Passed. To do that, you would have to have a regular USDA approved meat packing plant. And while I'll look 'em over for you and show you how we inspect them in the packing plant, you won't be able to say they are really inspected." It would have been hard to refuse his father-in-law and his friends, so he gave them a regular inspection, examining cervical lymph nodes, palpating the viscera, and finally checking out the split carcass halves.

Except for a very few old ascarid scars on one liver the pigs were nearly perfect. He was surprised, and said as much. Grandfather Hershfeld explained that he kept them on pastures not being used by his cattle, and that he moved them every few weeks, according to a pattern that insured that the worm eggs hatched out when there were no hogs around to infect. In an era when there were few effective drugs against worms, it seems that there were good farming practices that were of major benefit. Perhaps it shouldn't have seemed surprising. Mr. Hershfeld had graduated in agriculture at Iowa State.

When it was all over, evening had come and it was time for supper. Baths were in order, and supper was late.

"Daddy, do they kill hogs like that at the factory in the city?" asked Timothy as they all sat around the supper table.

"Well, almost. Except that they don't shoot them first."

"You mean they just cut their throats with them alive and kicking?" asked his wife, Elsie.

He had said the wrong thing, he realized. "I'm sorry to say they do. I hope that someday they will do better." After all, they did stun cattle, not because it was kinder but because it was safer.

Timothy spoke up and asked, "Wouldn't it be kinder to let them die of old age or something first?"

At that, Smith dropped his coffee cup. He was sure that there were people in the industry who would certainly endorse the idea but he could not explain it to all those seated around the table with him. At least the dropped cup changed the topic of conversation. It was not the first time Timothy had surprised them at the supper table with his concern for animals. Sometimes it got him into minor trouble.

A few weeks before, his grandfather's female fox terrier had produced pups. Timothy had wanted to know more than they had thought proper for a small boy to know about how and why puppies and babies come into the world.

It happened that his parents had come to visit and Mr. Hershfeld suggested that it was time to dock the puppies' tails. Naturally since he had a son-in-law who was a veterinarian the matter was raised at the supper table.

"You mean you are going to cut off the puppies tails?" asked Timothy incredulously. "Won't that hurt a lot?" It was less a question than a statement of fact.

"No," said his grandfather. "They are too small to really feel it. You don't remember when your tail was cut off do you?" Actually it was something else that had been cut off, rather than his tail, but the principle should have applied.

"You cut my tail off?" cried Timothy, greatly concerned.

His father couldn't resist.

"Of course we had your tail cut off. I had my tail cut off. Your mother had her tail cut off. Grandfather had his tail cut off. Decent people all have their tails cut off. Isn't that right sir?" He looked at his father in law.

Grandfather Hershfeld nodded solemnly, adding, "Otherwise we would look just like all the other monkeys at the zoo. Disgraceful!"

By now Timothy's face had turned bright red and he was really crying.

His mother said that perhaps she should put Timothy to bed and tell him a better bedtime story. With that, she took Timothy off to bed.

They were scarcely out of earshot when Grandmother Hershfeld said, "You two ought to be ashamed of yourselves. Filling little children's heads with such nonsense!"

Both men tried to say they were sorry, but she knew that they weren't. After all they knew it was part of growing up. Neither could they wear the properly contrite expressions on their faces.

Weeks passed. The attack on Pearl Harbor came and with America in the war there was much more work. There was a sense of urgency not previously felt.

Again Wilber was dreaming. He did not dream as much as before. Too often he simply fell dead asleep, only to be awakened by the alarm clock at five in the morning. Once in a long while he had fallen asleep in the office between jobs around the plant, or after his shift was over, or more likely between shifts.

It was their wedding day. There was a huge cathedral. Everything was splendid and beautiful. The ceremony had just ended and Wilber and Elsie were walking out of the church through massive doors on the way to their honeymoon trip. Inspectors lined the walk out to the waiting car. Resplendent in white uniforms, their scabbards and badges glistened in the bright sunshine. Their knives and steels formed an arch under which Wilber and Elsie walked. Church bells pealed. And pealed. Then somehow the sun was too bright.

It was five AM and the infernal alarm clock had just gone off. Elsie had turned on the light in his face. It was again time to get up and go off to work.

Some nights he would not even come home. He had eighteen hour days. He was never sure just which shift he might be put on. It was common for him to work two shifts. It meant more pay, lots more pay, but it was a great strain.

One of the products they were making was a canned meat product for the USSR that contained a great deal of fat. He could scarcely imagine how anyone could eat something that was almost like lard. He was told that even a winter in Chicago did not compare with those in Russia and that in such climates people could eat pure fat and like it. Eskimos did; it was a matter of habit and culture. So he thought of how much better off he was than say someone in Leningrad or Stalingrad and did his work cheerfully.

Smith wondered if the military procurement officers who often visited the plant had it any better than he did, since he had been deferred as a husband and father, and was serving in an essential government post. Indeed, he had a commission as a cavalry lieutenant as a result of having graduated from Texas A & M prior to veterinary school.

Had he been eager to fight, he might have found himself with General Patton. But he had heard enough about war not to be really eager to go fight the Germans or Japanese. It was not so much a matter of fear as doubt that there was anything he could do more useful than what he was doing. God knows, he thought, it was hard enough, and sometimes dangerous, even though no one was shooting at him.

One of his young veterinary coworkers had quit his job in the Bureau of Animal Industry as a meat inspector for active military service, only to turn up a few weeks later in the same plant as a procurement officer, at less pay than he had been making before and no closer to the shooting war. Except he couldn't quit for military service. He was already there.

The drudgery went on for several months. Smith did get some time off and could go out to the farm to see Timothy. His wife had gone into a good dental practice with two other dentists specializing in children's dentistry.

One night Elsie was feeling especially neglected, for Wilber had gotten up and was sitting in the old overstuffed Victorian chair looking out the window. She got up. She thought of beating him over the head with her breasts, or putting her hand over his eyes and saying, "Guess who?"

She put her hand on his forehead. He was running a fever and said that he felt chilled even wrapped up in a blanket. He coughed. He said he had to get up since he couldn't seem to get his breath lying down. They both knew what that meant. Fluid was accumulating in his lungs. It hurt to breathe. He had pneumonia. Cold, inadequate sleep, and fatigue had finally gotten to him. Before antibiotics and sulfa drugs, pneumonias were very serious, often fatal.

She dialed for a physician she knew through her dental practice. She had already taken his temperature. It was 104.

"That's not very high," he thought, "for a hog."

She read his thoughts and said, "That's much too high for someone your age." He was thirty-four.

The physician came. He gave Wilber some sulfa tablets and told her to get him to the hospital as soon as possible. He would have to be monitored. It seemed incongruous that something which he knew could kill him was not very painful. He didn't feel that sick. Except for not being able to take a good deep breath and having a slight headache he didn't feel too bad. Yet he had seen hundreds, if not thousands, of pneumonic lungs in animals and knew how serious it could be. He knew that it could kill in a matter of a few days. Dimly, he was aware that there was a specific human (primate) pneumococcus that was especially deadly. Much work had been done to develop sera to treat human pneumonia, with limited success. But, on the other hand, he also had seen many animal lungs that indicated that the animals had recovered from pneumonia, sometimes with no treatment.

At the hospital, he would have time to rest, to sleep, to recuperate. A new drug was used on him, penicillin. It seemed to bring him around to nearly normal in a matter of days. He thought that it would certainly be great to be able to use sulfas and penicillin in veterinary medicine. And so it was for a while, for perhaps a decade.

He did not realize that in time both of these drugs would pose great problems for meat inspection. The use of antibiotics by those engaged in raising farm animals eventually resulted in selected strains of bacteria which were resistant to those same sulfa drugs and antibiotics. Some very few persons were allergic to antibiotics like penicillin and could have very severe reactions to even the very tiny amounts that might remain in meat as result of the animals having been treated.

Some drugs, sulfas in particular, could get into the soil and plague even the farmer who had quit using them.

However, it would fall to those who followed Smith to deal with these problems.

The personnel director let him go on sick leave for a few days out to the farm. The war was winding down. Germany would fall to the allies later the following spring.

He and Elsie had managed to save a lot of money during the war. Both of them were working and with neither having any time to spend their money, nor much to spend it on, it accumulated. Much of it went into the stock market, as well as war bonds and the purchase of more farm land near the Hershfeld farm.

It seemed that Timothy was growing faster than they realized. Now he was eight. He could do useful chores around the farm. Both Elsie and Wilber felt that the farm was a more wholesome place than Chicago for a youngster to grow up.

Timothy had been given a runt female pig for a pet. He had named her Susie. With his care, she had grown into a good sow. To the surprise of the Hershfelds, she produced and raised several litters of pigs by the time Timothy was entering junior high school.

One day Mr. Hershfeld was working on a small low shed that he could drag around with the tractor. Old Susie the sow was keeping him company while he worked. Now Susie had a litter of piglets in the small patch of woods nearby. The shed was intended for Susie to have her pigs in next time. It was not likely that she would move this litter.

He happened to back into a sharp nail and in straightening up hit his head hard on the supporting beam. It knocked him quite unconscious. He awakened to find himself being dragged along by his pant's leg. He was wearing bib overalls. Far behind him were the sow shed and his straw hat. Doubled up under him was his other leg. Ahead of him was old Susie, his pant's leg held firmly in her mouth as she dragged him along. And farther ahead was the small patch of woods where the piglets lived.

It took a few moments for all of this to sink in; to assess the meaning of all this. After all, his head hurt, his backside in contact with the dirt hurt..., and...

He came to with a start. Susie dropped him with an embarrassed "whoof." He kicked at Susie and then staggered to his feet, growing angrier by the moment. He leveled an accusing finger at Susie and told her, "You were going to feed me to your pigs."

He stalked back toward the house, with Susie close behind. Every so often he would turn and kick at her and she would back off. Somehow she seemed quite contrite. She understood that she had done something wrong and that her friend and master was provoked with her. Surely they could come to some sort of understanding with proper apologies.

Then she did something odd. She trotted back and picked up his hat and again began to follow him.

He climbed the steps to the kitchen. He opened the door, and without a word to his wife who was making lunch, headed for the telephone.

"Stop right there!" She commanded. "Where on earth have you been? Just look at yourself! You look like you have been wallowing in the pen with the pigs, Ezra. Take off those dirty clothes right there. For goodness sake, your head's bleeding. Tell me what happened, and let me have a look at your scalp and that knot on your head. Maybe I should call the doctor."

He sat down, having stripped to his underwear. She put his clothes on the porch holding them at arm's length as she did so. She looked down and told him, "You left your new straw hat at the bottom of the steps."

"No I didn't," he replied, "it fell off when I hit my head." He felt the knot on his head.

"Sit down and let me put some iodine on it."

He did as he was told.

"Ouch!" The alcohol in the iodine solution stung. "Now I have to make that telephone call."

"What telephone call?" she asked, "You don't need a doctor. You need a shower."

"About Susie!" He exclaimed.

"She's right there, along with your hat," said his wife.

He got up and, squinting hard at the phone number written large on a paper pad on the wall beside the telephone, started to dial a number. He told his wife, "I'm going to call the packing house, and take old Susie there and put her in the freezer and we'll eat **_her_**...instead of the other way around."

Then he hesitated. He looked out the door and could see Susie waiting at the bottom of the kitchen steps. His straw hat lay beside her where she had placed it.

"She thought I was dead and she was going to eat me. She was carrying me off to the woods to feed me to her pigs..." he hesitated..." of course we have eaten some of her pigs... and she was going to share me... and she did think I was already dead."

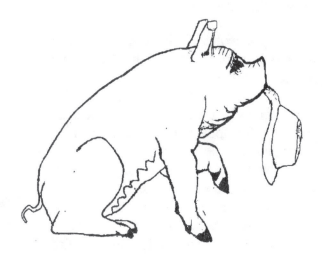

Slowly he put the earpiece back into its cradle. "I guess she isn't any worse than we are. I'll keep her, but I'll certainly have to keep an eye on her."

     With spring and summer came victory, first against Germany, then against Japan. Wilber and Elsie had time to see Chicago, the city they had lived in so many years and had yet to see together. They took Timothy to the sights she had seen many years earlier, before her marriage. There were the zoos, Lincoln Park and Brookfield. There was the Field Museum, Wrigley field, and much more.

     Their investments had been wise and they had the money to travel a little. They even went down to Texas to see Wilber's parents whom they had not seen since the wedding. And it was the first time Timothy had ever met his other grandparents.

America, America, the Promised Land

"Twice Hitler made me a widow and now I'm afraid it might be three times."

The speaker was Dr. Anna Wallace. She was a government veterinarian who had come to this country as the wife of an American Army officer. She was speaking to a fellow government veterinarian, Dr. Heinz Kruger. Dr. Kruger had been a German army veterinarian during the second World War but had come to this country afterwards. Before the war he had been a meat inspector in Germany. When the opportunity came, he joined the meat inspection service here in America.

Dr. Wallace had told her story numerous times to her comrades who shared her office, each time remembering something different, each time a bit more surprising. No one doubted the truth of her tale for those who knew her understood that she was an inherently truthful and exact person. It was part of her professional training. Unfortunately, she was often more truthful than tactful.

She had been Dr. Anna Paluski, then Kozincky, then Stein and finally Mrs. Stanley Wallace.

Of course her husband's name wasn't originally Stanley Wallace but Stanislaw Walewsky. He was a Polish American. His parents had changed their name a few years after settling in Hamtrammac where he was born. He had grown up speaking Polish as well as English. He had joined the Army during the depression, worked his way through college, gotten a reserve commission, and finally gotten back into the army as the war started with the express purpose of fighting Hitler. He had been married briefly in his youth but the marriage had failed and been annulled.

"Before the war," she continued, " before I was married, I went to veterinary college in Germany. There I met the man who was to become my second husband,

Gerhard Stein. However, he studied to be an engineer rather than a veterinarian. I had no romantic interest in him at the time we were students. When I met him again many years had passed.

"My first husband was a Polish air force officer, a land holder like my own family. I went to school to learn how to raise horses and cattle. I had no idea of doing anything like this." She waved her arms in an expansive gesture that took in the whole meat industry and all its works.

"We had been married only a short time when the war broke out and my first husband was killed. He was a flier and our old aircraft were no match for those of the Germans. Not long after that I again encountered Gerhard and we were married. He was my second husband.

"Or rather I should say, he had searched me out as a friend at no little risk to himself. I found that he hated the Nazis. Perhaps that is why I married him. I saw little of him for he was mostly in Russia. Then one day, I was arrested. I learned he had been arrested by the Gestapo. He had taken part in the Stauffenberg affair. He was one of those who had tried to kill Hitler. I believe they shot him. They tried to say he had killed himself, but I doubt it, because he was a devout Catholic.

"Then they eventually came and got me. First I was questioned, and then finally sent to Ravensbruk. The second American I saw when we were liberated was Stan. A few weeks later we were married. I really want to keep this one."

Major Wallace had been wounded in the latter part of the war and every so often bits of shrapnel would work its way into vital places and would have to be removed. Now was one of those times. The prospect of something going wrong frightened her.

111

"So I have had three husbands who were heroes, but I would have been happier to have had only one husband and have lived in simpler times. Being in Ravensbruk, especially because of Gerhard's part in the attempt on Hitler, caused Major Wallace to be interested in me. Unfortunately, I could tell him no more than I could tell the Gestapo for I knew nothing. I had not even seen Gerhard for over a year before his arrest."

Dr. Wallace was a pleasant middle aged woman, very large, very strong, and well, to put it bluntly, fat. She had reddish blond hair, freckles, and a huge amorphous area combining bosom and tummy.

Late in life she and her American husband, Major Wallace, had two children, a boy and a girl. After they had reached school age she had gone back to work. Rather than merely live in the Polish colony in Chicago, she had decided to work in the government.

"You know, I wonder if I ever met your husband, that is, your German husband," remarked Dr. Kruger. "Probably not, for Russia is a huge place, and there were very many men in the German Army," he hesitated, then added, "but who knows?"

They were both relative newcomers to America, part of a wave of refugees following the second World War. America needed them, both for their thorough knowledge of veterinary medicine and their willingness to work in circumstances that many American veterinarians found oppressive, and uninteresting. In a way, they resembled the veterinarians who went into government service during the great depression. Most had excellent general educations. They could read and write other languages and sometimes were called upon to translate documents the natives could not read.

There was an additional benefit. They were familiar with foreign diseases. For example, Dr. Wallace or Dr.

Kruger would have recognized Teschen's disease or hoof and mouth disease as quickly as most native American veterinarians would recognize hog cholera. Or at least they would have then, for eventually hog cholera was eradicated in the United States during the decade of 1970-1980.

There was, and always is, the fear that foreign or exotic diseases would make their way to our country. The fear was, and is, driven by the possibility that biological warfare fit the pattern of the cold war and clandestine war. The international terrorism that became pervasive later also increased the fear. Biological warfare well fits the plans and ambitions of brutal third world dictators. It is a logical extension of the terrorism they use and is readily available. Yet, third world populations are those most vulnerable.

We are protected by modern sewer and water systems, refrigeration, and sanitary packaging. All of these are developments of engineers rather than biologists, or medical men.

There is more.

Modern means of record keeping, telephones, and computers mean that trouble can be recognized quickly. Rapid transportation brings investigators onto the scene quickly. We would like to believe that the great plagues of the past are behind us.

However, as this is written, human cholera caused by Vibrio comma is spreading in the backward parts of the world. Acquired immune deficiency has appeared and is spreading. The tick borne Lyme disease is new and is spreading. Legionaries' disease arose and led to the discovery of a new group of organisms.

Yet even in the developed part of the world, there is something lacking. That is the will to take the necessary measures to remain safe.

There are included populations that do not live in the same world as the rest of us. There are groups that lack the

113

will, knowledge, or funds to observe all the rules of sanitation.

One can imagine the spread of contagious disease as being like the spread of a fire. If one scatters leaves on barren ground and then sets one of the leaves afire, whether the fire spreads will depend upon how close the leaves are to each other, whether the leaves are wet or dry, whether there is wind moving the leaves or the sparks coming from the leaves. And the fire must transfer from the burning leaf to another leaf before it goes out. Of course, embers may remain which wind may fan into a new fire.

Organic contagious disease lives and spreads in the same way. In this same sense, fire is the simplest form of life. Fire is self reproducing. It needs food, the fuel, and it must either move to the food or move the food to itself.

Population density is another way saying how close the leaves are to each other. Immune status is whether the leaves are wet, damp, or dry. The sparks are the disease vectors such as the mosquitoes that carry malaria. The embers are hidden carrier states as when the individual carries the organism but remains healthy and not contagious until weakened by other problems.

Yet even an individual who may seem to be in good health may transmit the infection to other susceptible creatures of the same or different species. Even people carry a mange mite which is dangerous to dogs but seems not to harm people; Demodex folliculorum.

The veterinarians and lay inspectors often discussed such matters. They talked about history and politics.

The two World Wars and the cold war was something most of them had experienced at very close hand, a great cataclysm, an epidemic of madness that for a time swallowed up their world. Perhaps talking about it helped. It seemed strange that the widow of a Polish officer would be talking to a former German officer who had once believed in the NSDAP (National Socialist German Workers Party); not just talking, but speaking fluent German. Yet Dr. Kruger summed up his feelings about his past beliefs when he said, "Next time I will know and recognize the Devil. He is a liar. That is how I will know! Things were bad for many people before the first war...but sometimes I daydream."

"What, Heinrich, do you daydream?" asked Dr. Wallace.

"I imagine it is that day in 1914. I am standing on that street corner in Sarajevo. I see the assassin's raised pistol pointing at the Grand Duke. I strike out and deflect the shot. The Duke and his wife return home safely. There is no first World War. My brother never goes to Verdun. Hitler becomes a successful architect, and I again become a veterinarian but this time a professor in the institute in Hamburg. Perhaps there is another world or universe where things happened in a happier way, but then again, there may be worse ones than this."

She said, "I am sorry all the bad things happened, but my life has been so complicated, especially since Stan and I have had children together. I don't think I would trade for a world without them. If I had not gone through all of that I could not have reached the happy place I am in now. I believe Stan will recover and we will go on."

"For you it is so, but for me..." Dr. Kruger trailed off. He had served on the Russian front during the war and his family had been killed in the bombing of Hamburg. His only son had been killed at Stralingrad.

When the day was over, she went back to the veterans' hospital to see her husband. Much better news came. The operation to remove the offending fragment would not be as dangerous as previously thought. Tomorrow they would remove it and all would be well. She thought about taking the day off, and then wondered what she would do with it. Their children were grown. They would be grandparents before long.

Soon her husband was returned to her, again fit and ready to go back to work in the office.

One day another trainee veterinarian was sent to their plant. The plant where they worked was used as something of a training ground for new veterinarians. Usually the government would try to send new people into good plants first to learn what things should be like. A little later they would be sent into plants that weren't so good so they could be shown problems and how to deal with them.

Dr. Julio Penario was of Cuban extraction on his father's side. His mother was Irish American. Sometimes the Cubans would make a slip and refer to her as an Anglo. She would bristle and let them know that she was Irish, not English.

His grandfather had left Cuba while Machados was taking over as dictator. Later there was Batista. There were cousins still in Cuba, even with Castro. With each new dictator more family members would come to the United States or Canada.

Julio was quite fluent in English which was better than the Spanish he had learned at home. After all, he was born in Miami. Had he not been interested in language and had a talent for speaking and writing, he might have lost his Spanish completely. But he had spent time in Puerto Rico and Mexico and made himself literate and fluent in both languages. He liked poetry and enjoyed attempting to translate it from one language to another.

Because of his exposure to other cultures, it was natural that he should have been interested in what the other older veterinarians had to say. There were long discussions, at least as long as time permitted, about freedom, liberty, equality, communism, and national socialism. Julio was sure that men ought to be free. Neither Dr. Wallace nor Dr.

116

Kruger was quite so sure of that. To some degree they believed in authoritarian, but necessarily honest government. Dr. Penario believed that authoritarian government could not be honest. To him, the notion of an honest dictatorship involved a natural contradiction of terms, an oxymoron.

One day they had a visit from the compliance and evaluation officer, a Mr. Kovacs. The compliance and evaluation officials were sometimes called the secret police. It was their job to see that the Inspection Service remained honest; that it was neither bribed, threatened, nor corrupted. They generally worked outside of plants tracking down problems like fake inspection brands, stolen meat, and sometimes financial skullduggery. If meat were condemned and then not accounted for, they were supposed to determine its fate.

Mr. Kovacs was a man with talent. To be a good compliance officer one must be naturally suspicious. If you were to write an aptitude test for compliance officers, one question might be;

Big flat rocks have hidden under them:
1. Nothing much
2. A few insects
3. a mouse or two
4. a snake
5. several large venomous snakes

A person with real aptitude will choose # 5.

Mr. Kovacs explained that when a company delivery truck had turned over during a snowstorm and then caught fire (and was a total loss) the meat and meat products carried by it had not been properly accounted for. The driver had been taken to the hospital, checked over and released, having only been shaken up. What had become of the truck's cargo? That was the purpose of his visit.

Not the company, but rather local citizens who applied marine salvage law, recovered most of the meat.

117

Mr. Kovacs wanted to be assured that the meat did not get into regular trade channels without reinspection. That is, any contaminated product had been brought back to the plant, condemned and destroyed. Any salvageable meat had to have been properly reinspected before getting back into commercial channels. It had been brought to his attention and he, Mr. Kovacs, required a full written report on the affair when he returned to the plant.

After Mr. Kovacs left, somehow Dr. Kruger and Dr. Wallace chose Julio to write the report explaining what had gone wrong and how such problems could be avoided in the future. He was not convinced that he should be the one to write it since he had not been there nor anywhere near the plant six months before when the incident happened. Dr. Kruger explained that most reports were written without first hand knowledge, but had to be written by people who were forced to make reasonable guesses. So he persuaded Dr. Julio Penario to write the report. He submitted his paper to Dr. Kruger.

Dr. Kruger read it aloud and then remarked, "No, it won't do. It's too simple. You will have to rewrite it."

Dr. Wallace demurred, "He has written the simple truth. What's wrong with that?"

"Everything. First, anyone can understand his answers. It is not written in Federalese. Look here. It says 'dirty meat'. That should be 'contaminated product'. Look at this! He said 'now' instead of 'at this point in time'. No wonder it is less than a page long.

"When I had to make reports to the Gestapo they were at least ten pages long if not longer.

"Young man," he said, turning to Julio, "when you make a report like this, you don't want it to be a model of clarity. You want it to be turgid, dense, and as dull and long as a Soviet novel. You aren't writing a good book. You don't want people to want to read it. You want it filed and forgotten. Even then you want it to appear to be something else so that it will seem to have been lost when someone goes through the files looking for it."

"My goodness, Heinrich!" she exclaimed.

"I understand your point, Dr. Kruger," declared Julio. "I will see what I can do."

And he did. The next report was voluminous, dull and very wordy as suits such a report. It went into great length about the social conditions of the neighborhood where the accident took place, explained that most of the "contaminated product, some 705 pounds of pork loins, 165 pounds of pork livers, 278 pounds of beef kidneys, were recovered, and denatured with an approved denaturant, and rendered in the rendering facility according to the regulations" citing chapter and verse of the Manual and the Regulations.

He cited figures for product brought back to the plant and reconditioned, figures for what had been on the truck when it left the plant, what was delivered before the accident, and finally a presumptive figure as to what disappeared immediately following the accident. To be sure, he took these from data given to the insurance company.

He said something about the conversion of delivery trucks except this one from gasoline to diesel engines pointing out that the fire that followed the overturning of the truck would not have occurred had the fuel been #2 diesel fuel. Such a conversion program was under way but had not reached this particular unlucky truck, etc., etc.

Had it been an armored truck carrying gold, he could have scarcely documented it more thoroughly. That the figures were all just figures and that the interviews had all been done by telephone mattered little. There was now a very impressive document which told far more about the matter than anyone would want to know if they were foolish enough to read the thing.

Dr. Kruger looked at the report, hefted it, thumbed through it, without really reading it.

"Splendid, Dr. Penario, splendid. You will make a public official yet!"

Mr. Kovacs was likewise suitably impressed. Like Dr. Kruger, he hefted the report, impressed with its weight and length, but unfortunately he sat down and began to read it.

"Where were you when all this happened?" he asked Dr. Penario.

"I was in the senior class at the University of Georgia getting ready for the state board examinations."

"Well then, how do you know what happened?"

"Why, I interviewed those involved whom I could locate; the driver, the policemen who investigated the accident and even the customers who were on the route both before and after the accident. And mostly I looked at the plant records. I looked for any discrepancies that might indicate that anything underhanded, other than the, uh," he hesitated to say theft, "unauthorized removal of the meat by persons unknown."

That seemed to settle the matter. Dr. Penario would learn to think like a bureaucrat and would stay with the service until his retirement some thirty years hence.

As Mr. Kovacs started to get up, his attention was attracted to a poster of a young lady without any clothes on but who had lines drawn on her naked body showing the various pork cuts. Mr. Kovacs stared at it long enough to set Dr. Penario and Dr. Kruger to wondering if there were some regulation against it. There probably was. It seemed to be the same look Dr. Wallace had given it the first time she had noticed it.

120

"Aha" said Mr. Kovacs, "a new study aid." He twisted his mustache. "That may get the inspectors' attention. After all we are meat inspectors!"

And with an appropriate leer on his face, he smiled, closed his books and left the office, headed for new adventures.

But we ARE
MEAT INSPECTORS

The Power of the Press, or Hell Hath No Fury Like a Woman Scorned

Olympia Westchester was the fashion reporter, editor of social news, and sometimes wrote as the food editor.

She had been with the paper (a Chicago daily now defunct whose name escapes me) for many years. Of course, Olympia Westchester was a pen name, for she had been born to the family of Moshe Baruk, shortly after the turn of the century. She had struggled to become educated in an environment hostile to higher education for women. She had struggled to escape grinding poverty, and had succeeded brilliantly. That her family relations had not brought her happiness was something that she accepted as regrettable but of lesser importance.

Felix, the editor in chief of the paper, had long wanted to get back at Olympia. She was vain. She was cantankerous. She was opinionated. And she was old enough to get away with it most of the time. What was worse, she had a loyal following of ladies who read her column in the paper and it was possible that the paper needed her more than she needed the paper.

Felix talked to the owners of the packing house and persuaded the Kowalski brothers to show the food editor around. After all, they had one of the best establishments in town. The brothers talked it over with their elderly father who was quick to point out that most housewives preferred not to know too much about where meat comes from. He quoted Bismarck on politics and sausage, explaining that both were best made out of sight.

But in the end they decided the good publicity would outweigh any risk so they agreed to allow the reporter into the plant. They trusted Felix.

Putting on her best smile, Olympia knocked at the door of the government office, and was admitted by Dr.

Smith who had been discussing the morning cattle kill with Mr. Kowalski and Dr. Wallace.

"Good morning, Madam. I'm Dr. Smith and this is Mr. Kowalski. What can we do for you?" Dr. Smith asked in his most gracious manner.

Mrs. Westchester answered, "I'm Olympia Westchester. We wanted to do a story on the meat industry and since our regular food reporter is indisposed I was sent in her place. I would like to be shown how meat is prepared here for the public. We especially want to get the real story of how the Government protects the consumer."

"Splendid," said Dr. Smith, with more gusto than real conviction. He had been informed about the proposed story by Mr. Kowalski and generally agreed with Bismarck.

"You will need some protective clothing, and..." he hesitated a moment, "you will need the permission of the plant management."

He looked over at Mr. Kowalski, who was the plant manager, as well as the kill floor foreman when his brother was absent.

"It's okay by us," declared Mr. Kowalski. He then asked her, "Have you got suitable clothes?"

"What do I need?" she asked. It was late spring and she was dressed simply but stylishly in slacks and a blouse.

Dr. Wallace spoke up and listed what Mrs. Westchester would need to tour the plant: boots, coat and helmet.

"I'll try to round up what you need," said Mr. Kowalski.

Mrs. Westchester had achieved a splendidly slender figure by exercise and diet. Looks were very important to her both professionally and personally. She had made numerous visits to the plastic surgeons, and had been reshaped a bit. The skin on her face seemed to have a mask-like quality.

123

That she had had several husbands was vindication that looks and charm counted and one never became too old for such things.

Mr. Kowalski soon returned with the necessary boots, helmet, and frock coat. They all were a bit too large, but they would do.

She stepped into the rest room and changed her clothes.

"Can you imagine looking like that?" remarked Mr. Kowalski to Dr. Wallace speaking quietly in Polish. "A strong wind could blow her away. She might have been pretty if she weren't so scrawny."

Dr. Wallace laughed softly and replied, "I did look like that when I was liberated from Ravensbruck in 1945. I vowed never to go hungry again."

They did not realize that Olympia could hear them speaking since her hearing had not been muffled by years of industrial noise; nor did they know that she had spoken Polish as well as Yiddish as a child. Unfortunately, she thought they had said she was as scrawny as if she had been in Ravensbruck. Naturally she was infuriated by the remark, but gave no outward sign that she had heard.

"Perhaps the best place to start would be at the beginning, out at the pens where the animals come in," remarked Dr. Smith. "She can watch antemortem inspection on cattle and hogs. Inspector, er, a, Bell is down there today."

A few minutes later, they were both down in the pens. She was able to stand back and see the animals driven off the trucks. She remembered the smells from her childhood when they lived in the part of town near the packing houses and stockyards. It hadn't improved any during the intervening years she decided. Hogs and cattle still stank. She wondered if the smell would get into her hair and what it would take to get it out again.

She was shown how the animals that had something wrong with them were separated and examined. Dr. Wallace was examining suspects, which at that particular point involved inserting a rectal thermometer into an old cachectic cow which was lying on the floor of the suspect pen.

The cow did not want to stand but a company man then persuaded her to stand up and get into the holding chute which was very narrow. The cow probably was infected with Johne's disease, which causes a chronic fetid diarrhea. It probably would not have affected Dr. Wallace's decision to condemn the animal, but it was good to have the animal's temperature on the record.

Dr. Wallace seized the cow's tail near the end with her left hand and then pulled it up over the cow's back. The cow did not like this and moved about trying to get her tail free. However Dr. Wallace had a very firm grip and soon the cow gave up and stood still. Then Dr. Wallace inserted the thermometer and held it so it would neither fall inside of the cow nor be expelled out onto the floor.

After a bit more than a minute, she spoke. "It's all right. You can come into the pen with me," said Dr. Wallace. "Here, so you can see better. See how emaciated the poor old thing is. She almost certainly has paratuberculosis. See how her feces is so black and foul smelling."

"No, thanks. I can see enough from here." Olympia thought she was going to be sick for a moment, but gritted her teeth and resolved that she would endure.

"Are you sure?" At that instant, Dr. Wallace removed the thermometer, and a shower of liquid cow manure shot out onto the pen floor. It spattered her white coat and boots. She then let go of the tail which was very dirty.

"It's all right. I will change before I go back on the kill floor," said the veterinarian, as she wiped the

thermometer off, read it, noting a slightly subnormal
temperature. "Incidentally, the disease is not transmissible
to people, so far as we know."

Then she washed off her hands, boots, and the
thermometer with a water hose which was nearby.

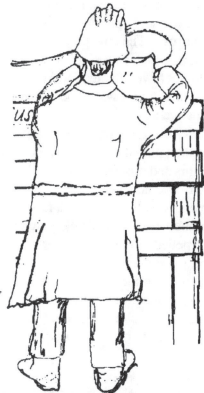

Inspector Bell tagged the cow in the ear with a red
U.S. Condemned tag. A company man came with a big
captive bolt pistol. He put the muzzle of the weapon on the
cow's forehead and squeezed the trigger. There was a
muffled report and the cow convulsed and then lay still.

A few minutes later, a tractor arrived to drag the dead
cow, along with some dead pigs, away to the rendering
facility. A company man slashed the carcasses with a knife
and then poured a green dye on them as required by the

Regulations before he dragged them away. Dr. Wallace explained to Olympia that this wasn't really necessary in a good place like Kowalski's but might be needed in some less honest meat establishments.

Dr. Wallace said, "Now let's follow them over to the tank house. You can see what goes on there. I can also examine the intestines of the cow to see whether I was right as to the cause of her being so emaciated. If she had Johne's disease the intestine will be much thickened, like a hose. She really was not such an old cow according to her teeth."

The intent look on Mrs. Westchester's face was not the result of her intense interest in pathology but rather her battle with her own viscera which were on the point of rebellion. It did not occur to Dr. Wallace that one might not really be interested in biology.

Leaving Inspector Bell, they followed the tractor to the rendering plant, a walk of perhaps a quarter mile. Here were different smells, different, but certainly as vile in their own way. There was a smell like grease boiling, with other foul smells thrown in for seasoning.

"How do you stand the smells, Dr. Wallace?" asked Olympia.

"One gets used to it; however I find it worse when I have been away from it, like when I come back from vacation. You may call me, Anna, if you like."

"All right, and you may call me Olympia. After all, my readers do."

Dr. Wallace gave a detailed explanation of how inspection protected both the human and animal population; how disease found in the meat packing establishment might be reported to the field service with government men going out to farms to stamp out disease; disease which might be dangerous to people as well as animals. She gave a short history of anthrax outbreaks in Europe.

From the doorway, Olympia watched as Dr. Wallace cut into the belly of the cow. She could have had the man who was in charge of the tank room do it but she wanted to see the cow intact and do her own dissection. Perhaps there was an element of showing off in it. Certainly she believed she could do as well as any man. The men who worked with her were quick to admit that she often surpassed them.

"There it is, the thickened intestinal wall, characteristic of Johne's disease. I was right," she said, holding up a short length of intestine. "There is a lot of edema due to emaciation. The disease starves the cow to death!"

"That's horrible!" muttered Olympia under her breath.

"Yes. Disease is often horrible," answered Anna.

"Can we go back to the office now? I would like to see the rest of the plant."

Her question was immediately answered by the blowing of the plant's steam whistle. It was lunch time, at 10:30 A. M. So they went back to the government office. Other inspectors were coming in to the office. They had been working the on kill floor and were very dirty and smelly.

As they came in, Dr. Smith introduced each inspector one by one. They took off their very dirty coats and hung their hard-hats and coats on hooks on the wall. She was surprised at how many they were; five from the kill floor, two from processing, and one from the pens. Sometimes they had a man at the tank house.

They had little to say and seemed to talk in grunts. She realized that they were bone tired. She had arrived at what was a reasonable hour for most businesses, 9 A. M. and found that the plant had begun operations at 5:30 A. M. when it was still dark.

"Will you have lunch with us? Did you bring your lunch?" asked Dr. Smith. "If not some of us will be going over to the Stockyard Cafe later. You notice we run on an earlier schedule than most of the rest of the world."

Olympia demurred. Something rather the opposite from eating was on her mind. "Perhaps later," she said.

One of the men stated that the cattle kill was over for the day and that just as soon as the clean up was complete, they would start on hogs. Of course, that would be after lunch. The plant could have killed both at the same time, but did not, mostly because of a lack of sufficient workers and space in other departments. Since a kill floor is one of the most expensive parts of a plant to build, and much of the rest of a plant is built around it, it is often larger than really necessary to allow for future expansion should business prosper.

"Too bad you won't be able to see the cattle kill today. Perhaps you can visit us again later and see it all. But you should see enough to have something interesting to write about by the time we finish today," said Dr. Smith.

"Perhaps," she answered, but she thought it would be a very long time before she made a return visit if she had anything to say about it.

The whistle blew again. Dr. Smith got Olympia a fresh clean coat and took her to the shackling pen where the hogs were run into the building. The noise was deafening, literally, with the hogs squealing and the machinery roaring.

The pigs were driven into a narrow chute and a man touched each hog on the back of the neck with an electric stunner, a long rod with electrodes on the end, like a pitchfork with two tines. He pushed a button and instantly the hog stiffened, with every muscle contracted violently. A moment later the hog was propelled onto a heavy metal table where another man put a chain around its hind leg. Up into the air went the pig. A third man cut its throat with a long knife. The hog began to twitch and shake, but made no sound. It disappeared into the depths of the building.

What shocked her was the methodical way the men worked, as though they were merely building a fence. Obviously what was a novel and horrible experience for the hog was routine and boring for them. Before the day was over, more than two thousand hogs would go up that chain. She wondered how killing could be so routine.

They went inside the building and circled around until they could see the scalding vat. Hogs were being dragged through the hot water. Sometimes they would seem to struggle as they went under as though the hot water could awaken them from death. The water was very dirty.

At the end was a huge machine which beat the hair from the hogs. Just beyond was another machine which played gas flames up and down on the hog carcass to burn off any remaining hair. Some of the hogs had black or gray skin, a few were slightly brown, but the vast majority were white. Like human skin, she thought, and then she wondered what the black men thought about the color of the scalded hogs which when the hair had been removed was as white and clear as hers.

Dr. Smith explained that the release of the hog hair depended upon a spinal reflex, so that the temperature of the water was critical down to the degree. He surprised her by dipping his hand into it very quickly showing that while it was hot it wasn't even close to boiling. He started to explain that a hog needed to "have some life in him to pick properly" and that a hog that had lain around dead would not bleed out nor shed its hair properly in the dehairer and then thought better of it. It wouldn't be fair to the Kowalskis to tell about some of the bad things that happened in other plants. If you can't say something nice don't say anything, he thought.

They continued. Workers swiftly trimmed feet and ears. This was a modern live rail system that pulled hogs through the entire plant by a series of overhead chains, with pulleys holding the hogs up onto an overhead rail.

An inspector was busy cutting lymph nodes on the hog heads looking for abscesses and tuberculosis. Every so often, he would find something that should not be there and drop the head into a tub labeled U. S. Condemned with a quick cut of his knife. The time had not yet come when tuberculosis found in the head would be noted and the head tagged to inform the other inspectors. Workers cut the heads off a bit further down the rail.

They walked further along. A man opened the carcass with a single quick long vertical cut, and then removed the entire viscera and placed it in a stainless steel pan, one of many, set into a sort of conveyer, pan beside pan, in a chain. The pans ran for perhaps forty feet, turned down into a hopper dumping the contents. Underneath on the return trip, they were washed, and returned from below, to carry another set of hog viscera.

The dumped viscera fell into a machine that ground them up and expelled them into a pipe which carried them into the tank house where they were dumped into a rendering vat.

"What is that?" she exclaimed. She had just seen a pan go by containing hog viscera. She had been watching the pans for a while but the viscera were normal except for livers with white scars on them caused by migrating ascarid larvae. Most of these were being stamped U.S. Condemned, and were going down into the chute which led to the rendering tank.

This set of viscera was different. The intestines seemed to be still alive and writhing. They had the appearance of an automobile wiring harness, tightly wrapped with tape. To answer her question, Dr. Smith reached over and picked up the entire set of viscera; lungs, heart, liver and intestines, and placed them into the examination pan directly in front of them. He named the parts for her, making a few cuts with his knife. "See the white spots on the liver. That is a really bad one. It even has abscesses on the liver."

She pointed to the intestines, which were moving. It wasn't her imagination. It was a horrifyingly real movement. She could hardly hear his explanation because of the ambient noise.

He cut into the intestine. Immediately out spilled adult ascarids, white shiny worms nearly a foot long and perhaps a quarter inch thick with pointed ends. The effect was that of an obscene bowl of spaghetti come to life.

She had to turn away. "I'll never be able to eat spaghetti again; and I'll certainly keep Kosher so far as pork is concerned," she thought.

"I've seen enough of this part of the plant. Can we go to the office? Then I'd like to see the rest of the plant," she said.

At that moment, a hog carcass was railed out for final inspection by the veterinarian and trimming by a company trimmer, who at this time happened to be Mr. Kowalski. One of the hind legs was far too large. There was a very large inguinal lymph node easily visible inside the body cavity next to the leg. Perhaps it should have been the size of a marble and white. It was the size and color of a plum. The viscera showed no particular abnormality. Mr. Kowalski and Dr. Smith looked at the hog.

Dr. Smith would have to examine the hog before going back to the office so he showed Mrs. Westchester how he examined a set of hog viscera with a few deft strokes of the knife. Then he examined the carcass, nodded to Mr. Kowalski who took a hook and some cord and began to tie the hog to the gambrel above.

"Why is the hog's leg so big?" she asked. The leg had been broken, where, when and how, they did not know. Perhaps it had been broken in transit, perhaps while the carcass was being pulled through the shackling pen or the dehairing machinery. When Mr. Kowalski cut into the leg there was a large blood clot and a broken femur. He cut the leg off and Dr. Smith took off the retained tag. He turned to Dr. Smith and said that he was glad it wasn't an abscess. It would be a matter of some small investigation between them for had the hog arrived with a broken leg, it should have been classed as a suspect and been brought in later.

Dr. Smith was glad it wasn't an abscess, too. That probably would have completely undone Mrs. Westchester. He suggested that perhaps the leg was broken after the hog was dead. He hoped so though he inwardly doubted it. The clot had been too large.

They went back to the office and got another clean coat for Mrs. Westchester and washed her boots and hard hat. Now she would be shown around the processing parts of the plant, including the sausage kitchen and the shipping floor. Perhaps this would be more congenial to her.

Another inspector took her around through the processing part of the plant, explaining the uses of each of the giant machines. He did not fail to explain how dangerous they could be if misused or handled carelessly. They observed two mechanics working on a giant mixer-grinder. He pointed out the padlock on the electric switch, explaining that there was only one key issued, the others were in an office safe. Only the senior mechanic had a key and could unlock the switch. The machine could not be started accidentally while being serviced.

There were matters of labels and packaging. The color red could not be used on transparent bacon wrappers. Only so much water could be added, and had to be declared, usually ten percent. Sometimes water could be added to carry in other added ingredients, but then had to be cooked off in a smoke house where the hot dogs and bologna were cooked.

To Olympia, it all seemed dreadfully complicated, rather like the Rabbinical law her cousin Mordacai had studied and tried to explain to her.

"If there is one thing we would like to leave you with," said the sausage foreman, a man whose knowledge resembled that of a chemist rather than a blue collar worker. "Tell your readers to read the label. We and the government make a great effort to declare the truth on our labels. It is the only way to be sure what you are getting."

He continued, "There are five standard items on a label; the name of the product, name and address of the company producing it, the establishment number, directions such as 'keep refrigerated' or 'cook thoroughly' and last but most important, a list of ingredients in descending order of predominance -- oh, and the net weight."

Her head swam.

He repeated sternly, "Read the label! Be sure to tell them to read the label carefully." He showed her one of the smoke houses, a huge room like a box with rails suspended from the ceiling, like the carcass coolers she had seen earlier. But it was very hot, 155 F, it said on an outside thermometer. He opened the doors, and a delicious odor filled the hallway. The inspector stepped forward and took a small thermometer from his coat pocket and stuck it into one of the hot dogs hanging in loops in their plastic casings. He did this several more times, nodded to the foreman in satisfaction, and signed his initials on a card, along with the time and date and recorded temperature.

The foreman cut off one of the hot dogs and peeled it out using a long pocket knife. He cut off a piece, and ate it, another he gave to the inspector who also ate his little piece of hot dog.

"Mighty good for an all beef hot dog," remarked the inspector.

"They will never be quite as good again," said the foreman.

Suddenly Olympia realized she was very hungry. He must have read her mind. He cut off another hot dog, partly peeled it and offered it to Olympia.

She hesitated only a moment, and then took a bite. It was delicious. She quickly finished it. "That is the real test. Put your mouth where your signature is," said the inspector.

"A person could get fat with such temptations about," she thought. Finally, it was all over and she was back in the office. If she had not seen it all, she surely had seen enough; more than enough, too much she thought. It was mind boggling.

She had already noted the girth of the Kowalski brothers, as well as that of Dr. Smith and Dr. Wallace. She remarked that working with and sampling food could be a problem.

"Well, you see, Mrs. Westchester; it's like this. All of us handle meat, especially pork, which is fat and greasy. We absorb the fat through our skin and eventually it is deposited on our tummies. Isn't that essentially what happens, Dr. Smith?" said Mr. Kowalski with a sincere expression on his broad Slavic face. "After all, frogs absorb water through their skins, and babies absorb food through their skins."

Dr. Wallace turned away so as not to be seen laughing. A moment later, with great composure, she and Dr. Smith nodded solemnly.

Olympia said nothing, as if the joke had eluded her. She suddenly remembered what had been said about Ravensbruck and how angry she had been.

It was a few days before her article appeared in the paper. Neither the Kowalskis nor Dr. Wallace, nor Dr. Smith had seen the article. How could they have seen a morning edition when they were at work when it was delivered? Eventually there was a mid morning phone call for Dr. Smith. On the other end was the district director who asked, "Have you seen what that lady wrote?"

"No," Smith answered, "Why?"

"You read it and tell me how she got that out of what you told her," and he hung up.

Hardly had he hung up than Steve Kowalski walked in with the morning paper and, shoving it in front of Dr. Smith, said, "Read this. That dumb broad doesn't even know a joke when she hears one. Papa was right!"

The headline read, "Epidemic Obesity Threatens Agricultural Department Inspectors...USDA Official Avers!"

The story began with, "Dr. Wilber Smith, a senior official of the United States Department of Agriculture today declared that handling fat meat caused obesity in meat inspection personnel and plant workers.

"However physicians and dietitians could not confirm this finding."

There was more about diseased and dying animals and bad smells. Finally there was this admonition, "See tomorrow's edition for more hard hitting expose of the meat business."

The four of them looked at each other and were silent for a long minute. Then each began to laugh, long and hard. It was a good joke...on them.

Anna Wallace caught her breath and remarked, "Just imagine how it would have been if our iconoclastic Dr. Kruger had been here instead of on leave. He would have really teased her." They all laughed again, even louder.

It turned out that no harm was done. The public is very forgetful and prefers not to think about where meat comes from. What is even stranger is that some years later Anna Wallace and Olympia Westchester became friends because of the efforts of mutual friends.

Still, even if no harm was done, neither was any good. The public got a poor explanation of what the meat industry was all about.

Pussy Cat, Pussy Cat, Where Have You Been

It was early morning and still dark outside. The ladies who worked in the sausage room had gone into the women's locker room and were getting dressed and ready for their day's work in the plant. Later, a few others would go onto the kill floor. Most would go into the processing parts of the plant. Many had very responsible jobs. After all, kitchen and cooking was woman's work. Only here, the scale of operations was much larger. They would be preparing enough meat products to feed a large city although in reality, their products would be shipped over most of the Midwest. Many were of grandmotherly age. They had always been there but now since the second World War, they were there in greater numbers.

The men were already down at the pens, and out on the kill floor. A few were still left in the men's locker room. Suddenly the men heard screaming, and some laughter coming from the lady's dressing room. Those men closest headed in that direction with visions of ax murders and other terrible things going on in there.

However, before they even got to the doorway, a huge rat ran past them, followed by a big white tomcat in hot pursuit. Inside, ladies were standing on benches and chairs, holding their skirts tightly about themselves.

It seems Angel, the cat, had caught a rat down in the animal pens and since it was such a splendid rat had brought it up into the lady's locker room to show off. However, when Angel put the rat down, it was far from dead, and picked itself up and ran. Pandemonium ensued. Ladies shrieked, climbed the benches, threw boots and shoes at the rat and yelled for help as loud as they could.

Cat and rat disappeared into the night. Whether the rat escaped and recovered we do not know.

Great bulls and boar hogs have run loose on kill floors without causing the disturbance a rat and a cat can cause, at least amongst the fair sex.

Angel was one of several plant cats, and was at the moment the reigning tomcat. Animals for slaughter must be fed if kept over from one day to the next, and especially over weekends. They eat grain. So do rats and mice. Rats also will eat meat and meat scraps when available. They are not housebroken so they are very unwelcome in food establishments of any kind. Rats somehow are able to dig long tunnels and stay out of sight and yet be around in large numbers.

Angel was reasonably housebroken, something rather unusual for a tomcat. So while regulations forbade him entry to the food producing parts of the plant, he was allowed into the offices, including the inspection office. Normally he lived outside down in the cattle pens.

"How much does a rat weigh?" asked one of the men working in the pens.

"Maybe a pound. Unless he's roosting on one of the scale beams or arms under the weighing scales. Then he might weigh a lot more, maybe fifty pounds," answered the weigh master. "Its a matter of leverage. Trouble is, he might be on the farmer's side of the beam and you might be weighing in a few fifty pound rats, along with the hogs, or steers. Terrible thing to be buying your own rats, over and over again."

He continued, "Each time we weigh in a bunch of animals we have to balance the scales. Its a matter of law, but besides, the first animals leave manure in the scale pen and that accumulates during the day. We don't want to pay for that twice so we clean it up, but...when you try to balance the scales and they seem to change weight with nobody on them then look out...for rats, or even a cat for that matter. Angel here could weigh as much as a cougar.

"Couldn't you Angel?" He reached down and stroked the cat. Angel was solid white with orange eyes. His fur was short and he was as muscular as a bulldog. He was big too. The only trouble was he was trusting and friendly and folks were afraid he might be stolen. Still, there would be another to replace him in a short time. Few cats survived more than two or three years around the plant. Although they always managed to get someone to feed them, the dangers around the plant from vehicles and animals meant their lives were usually short. Hogs would kill and eat cats if they could catch them.

Dr. Kruger had taken a liking to Angel some time before when he and Dr. Wallace had been working the plant. He had described a similar cat catching rats and bringing them to a group of entrapped soldiers in Stalingrad. They would dress out the rats and eat them, giving the viscera to the cat. "Rather like rabbit or squirrel," he remarked. Finally the cat failed to return. Perhaps it too had fallen victim and been eaten. He explained how cats could be distinguished by the different shape of the scapula from rabbits, once the pelt, head, and feet were removed. After all, in Asia they were food, too.

"What did the rats eat?" innocently asked an inspector.

"Better you shouldn't ask!" Dr. Kruger replied grimly.

"Did you ever eat a cat?" asked another inspector.

"I hope not," he would say, looking at Angel.

Dr. Kruger was soon transferred to a small plant in Wisconsin where he would be near his older sister. There he spent the rest of his career.

One morning early before the kill began, Dr. Smith and Mr. Kowalski had hardly settled into their chairs when another man came in. He was obviously another veterinary inspector with his white USDA helmet, badge, and clipboard.

"I'm Dr. Mullins and I'm supposed to do a brief review of your plant this morning. I take it you knew we were doing some unannounced reviews in the area."

It was true. The word had gone out to expect such reviews. Dr. Smith had thought the idea a good one in the abstract, but was not so sure now that a reviewer was on his door step. The time would come when he would be doing some reviews himself.

"First, let's see your brand log," asked Dr. Mullins.

Dr. Smith opened the brand box with his key and showed Dr. Mullins the actual brands still in the box, and the brand inventory showing how many of each kind of brand were in the plant. Oddly, the brands were bought and paid for by the plant, but were treated like government property, logged in and out, signed for, and if lost, the loss had to be reported immediately. The brands came in three sizes, and were especially made with the number of the establishment on them. The brands tallied correctly. They were either in the box or assigned to the plant for use at the time. Next came the inventory of seals. These were metal strips, each with an individual number, a long number, in series used to seal certain shipments, from one Federal plant to another.

Then came the files. Files were supposed to be in a defined order, identical from plant to plant, in their order but different in that plants were different. One started with the Grant of Inspection documents, a sort of charter for the

141

plant. There would then be the water report indicating that the water was tested within the prescribed period and found safe, six months if from a private well, a year if from a city water supply.

Finally, as the reports were picked through, one finally came to labels. All sorts of things can be wrong with labels, arcane minute details. But fortunately Dr. Mullins was a kill floor vet like Smith so he did not spend much time in the files. He wanted to observe the operations.

He brought something else new to the review...a light meter.

"A what!" exclaimed those surrounding him. They were all staring at the small object he held in his hand.

"Regulations now require fifty foot candles on all working surfaces, about the amount of light cast by a 150 watt bulb at three feet above the working surface," explained Dr. Mullins.

They started inside rather than at the pens, as was usual. Near the boilers and the tank house. Perhaps something was on his mind that would eventually reveal itself.

The pens were a bit dim, not quite meeting the required 10 foot candles. However, as if to make up for it, there were three hundred foot candles shining down on the weigh master's desk.

At the tank house Dr. Mullins bent down, examined a hole in the wall carefully, which led into the kill floor near the dehairing machines. He plucked a single white hair from the edge of the hole and held it up to the light.

"American shorthair male, about a year and a half old," he remarked as he studied the single hair. Mr. Kowalski and Dr. Smith looked at each other and said nothing.

"A what?" exclaimed Mike Kowalski.

"A tomcat," replied Dr. Mullins.

142

"How did you know that?"

Dr. Mullins winked at Dr. Smith and said, "Well, a veterinarian is supposed to know such things...besides I saw him come through the hole on the other side when you were looking the other way. Anyway, this hole should be closed up. If you need it for pulling a hose through, then line it so it's smooth and put a cap or cover over it."

They went on to the kill floor. Before long the killing of hogs would begin for the day. There in the middle sat Angel, licking a forepaw and looking totally bored with the whole thing. Reviews were nothing in his young life.

Clean white outer Garments

"I don't believe the regulations allow a cat on the kill floor, even when the kill is not in progress."

And of course they don't, as all three of them knew very well.

"However, we can credit you with having a professional rodent exterminator and I see your employee has clean white outer garments," Dr. Mullins scratched down some notes on his clip board.

"Put in some more light in the pens, too."
When it was all over, they hadn't done too badly. Dr. Mullins had encountered a rat in the spice room in another plant a few days earlier. He remarked to the plant manager accompanying him, "Cute little feller, what's his name?" before lowering the boom on the plant.

Rodent infestation requires a drastic cleanup, along with condemnation of exposed product. What was worse was that in this case it turned out that the rat had been captured by some disgruntled workers and saved especially for the occasion. We found a drawer with food, water, and rat droppings during the cleanup.

Cute little Fellow,
What's his name?

"Maybe you can explain something to us that we don't quite understand," said one of the Kowalski brothers addressing Dr. Mullins.

It seemed safe to ask questions now that the review was over. They were all standing around on break in the government office. Dr. Mullins lit a cigar. He remarked that it was a twenty mile cigar. That is, it was of such size that he could finish it by the time he had driven to his next plant which was twenty miles away.

"Well, I'll see if I can explain it. What is it?"

"Up until maybe last year, if a hog had TB we just went on and passed it unless there were what Doc called systemic effects. Not like cattle with TB. Now we can let the carcass go if it's just in the head (which is condemned), cook the carcass at 170 F if it's in the head and the guts, excuse me, visceral lymph nodes, but if it's under the liver or in three places, it's all condemned, not just the head. Somehow it doesn't make sense. If it ought to be cooked for two places, then why not for one, or three for that matter."

"Unless it really is bad," remarked the other brother. He meant "really looks bad."

"I've got to agree with these folks," remarked Dr. Smith. "It doesn't seem to make much sense medically speaking. If cooking will really kill TB then shouldn't they all be cooked?

"I even had an inspector over in one of the small plants where they inspected the viscera first run into a set that had it so bad in the hepatic lymph node he condemned the carcass and then couldn't find any in the head. The plant foreman said the carcass shouldn't have even been examined and the vet, well, I said it should be condemned, which we did. Fortunately the plant people were pretty good about such things and were just having fun with us.

"I guess you knew we had so many there at first we were letting lay inspectors condemn for TB in remote plants."

Dr. Mullins drew a deep draught on his cigar, rolled his eyes back into his head and thought a few moments. Then he said, "I guess I'll have to level with you people.

"I'll admit that medically speaking, all you say is true. It doesn't make much sense. Either it is or it isn't. However, all that we do has to make sense in another way."

145

"What's that?" asked Steven Kowalski.

"It has to make political sense. I assume all of us here would like to get rid of swine tuberculosis. It certainly doesn't help the swine industry, doesn't help the meat industry, and we sure hope the public doesn't get excited and confuse bovine or human TB with avian and swine TB"

"No!" said the Kowalskis, almost in unison.

"If we had started on day one condemning all hogs with any TB, or even requiring that they be cooked, the farmer and the packer would have raised such a fuss because of the money lost, we would have had to drop the rule the first week. So we decided to cut off the dog's tail by inches, a little at a time. When we get the incidence down enough maybe we will be able to simply cook or condemn all of them. We had to start somewhere if we were ever to get the job done."

"What do you mean the farmer?" remarked Mike Kowalski. "We have to pay for the hogs."

"Yes, but we have a long memory and if a farmer brings us a bunch of TB hogs, we won't accept them at the regular price any more," stated his brother.

"Well, we will just have to tell the farmer not to feed uncooked chickens to his pigs and not to let birds crap into the hog feeders. Simple enough, isn't it?" commented Dr. Smith.

"Well, gentlemen, I must go," declared Dr. Mullins. "You have a pretty good plant. Give the cat a pat on the head from me." And with that Dr. Mullins left the government office carrying his briefcase, helmet, white coat, and boots in hand.

As he walked towards his car, Mike Kowalski watched from the open doorway and remarked to the others, "That sounds fine but, I'll bet there is more to the TB than that. It's not going to just roll over and play dead that easily. What do you think, Dr. Smith?"

"Well, we won't know until we try."

"You know, it's the farmer who will have to try. It's political all right. It's the farmer who will have to do the job. He raises the hogs. We just buy them and make them into pork. If we raised our own hogs like the poultry people raise and slaughter chickens we could do it."

"Look Steve," said his brother, "there are a lot more farmers out there than there are of us. They vote. The farmer is a good guy and we are the bad guys. So the government bashes us and we in turn bash the farmer. They might want to bash the farmer but they don't have the votes. And speaking for both farmers and packers, I sure hope the public doesn't get into the act. You remember the cranberry thing a few years ago. We really ought to have enough sense to solve our problems ourselves."

Yet despite serious efforts by hog farmers over the years to prevent bird to hog contact, swine TB persists, although at a greatly reduced rate. There have been a number of less than satisfactory explanations, at least unsatisfactory in that they do not offer a quick and easy route to eradication of swine TB Still, it was an effective beginning.

Angel was removed from the plant but for reasons you might not have expected. Mike Kowalski's wife raised show cats, Siamese cats to be exact. One day she came down to the plant and saw Angel. "A perfect American short hair," she exclaimed. "This no place for a fine cat like you!" And she took him home with her. She fed him. She washed him and brushed him.

She remembered that at the particular time some American Shorthair cats were allowed into the registry and the show circuit without their particulars being known, a situation which did not last long, but she made the most of it. Soon Angel was a Grand Champion, an honor her Siamese cats had never won. So Angel became a house pet.

147

Eventually the plant had another reigning tomcat. This one however was solid black and bore the name, Soul Brother. Soul was long on street smarts, rough on rats, and for all we know may still be on the payroll at Kowalskis'. They were, of course, equal opportunity employers.

The night after the review Dr. Smith had another dream. He was reviewing Old Nasty. He was carrying a huge flashlight and was accompanied by a retinue of other inspectors. One of whom carried a light meter, another carried a clipboard. He read Farkas and Weissmann their rights, as was their due, and then began to inspect the plant. There was only one light in the plant. It was directly over the cash register.

There was a chorus singing:

"The light is dim

I cannot see,

I have not brought my specs with me."

Meanwhile a very official voice from above was reading the regulations prohibiting each and every offense known to meat inspection.

When it was over, the inspector secretary was carrying a huge pad and clipboard; in fact staggering under the load.

The indictments were read and sentence passed. The miscreants were stood against a wall and washed with kill floor hoses, though not in 180 F water.

Huge machines, bulldozers fired up their engines, with a thunderous roar. They headed for Old Nasty and began to tear it down. The roar dissolved in the ringing of the alarm clock. It was time for another day.

Neither Jew nor Greek

It was a bright cold day in April. Easter would come in a few days. That is, Easter would come for Dr. Demitrious Christophoros and other Orthodox Christians. Because of that, he and his assistant, Inspector Jones were standing in a pen surrounded by sheep.

Lambs, really, for they were young enough to still qualify as lambs with the epiphyseal caps on their long bones still unossified. That is, the cartilage at the ends of the leg bones had not yet turned to bone. They were really a nuisance for they were too tame to handle and drive well. They crowded around like pet dogs about to be fed and had to be pushed aside. Parading them to and fro for a proper Antemortem examination according to the Manual of Procedures was a problem.

It had not been very long since the Jewish Passover which had to precede Easter, when there had been more sheep.

"There goes old Stonebagel," remarked Jones, who had noticed Rabbi Bekelstein, the schlacter, headed for the kill floor. Rabbi Bekelstein was the man who killed the sheep and cattle according to Hebrew ritual dietary law.

"How can he just cut their throats like that?" remarked Jones, mostly to himself for he knew full well how it was done. Humane slaughter required that, except for Kosher killing, animals must be rendered unconscious before they are bled to death. He remembered that not many years before, hogs were simply hoisted up and their throats were cut without any thought that it might hurt - the hog, that is.

"Why don't you ask him?" said Dr. Christophoros, a bit of grim humor in his voice. "I'm sure he would tell you."

But Jones already knew what the rabbi would say and had said. It had been a matter of discussion between Dr. Christophoros and Rabbi Bekelstein many times before.

"It is required that the animal be conscious, according to the Law. I did not make the Law. I must obey it," said the rabbi.

Which was almost exactly what the good doctor often said, although they did not refer to the same law. Actually, Dr. Christophoros referred to the Regulations. There was one point upon which the rabbi could hang many arguments. Regulations were made in Washington, and were often changed, while the Law was made in Heaven and did not change.

It was a sort of proverbial saying around the plant that while the doctor and the rabbi were opponents on many occasions, they were friends, rather like a dog and a cat living in the same household. They both were well educated, liked to argue (although they called it a discussion), and often agreed on many things (though they would usually deny that they often agreed).

They had reached a point at which both could remark that they were "both Orthodox, however..." and "I'm merely a cantor, he is the rabbi," (or vice versa) with some humor. If people were left in confusion at times, it was their problem. Neither Rabbi Yehudi Bekelstein nor Dr. Demitrious Christophoros concerned himself about it.

Both were dark swarthy men of medium height. The veterinarian had a large mustache. The rabbi had a full beard. Both were slightly bald. Finally, both men had been born in this country shortly after the first World War to parents who had come from another country. They shared three thousand years of Mediterranean civilization.

On this day, only a few of the sheep were intended for the Kosher trade, the rest would go for the Eastern Orthodox Christian community; Greeks, Russians, Armenians, Romanians, and Bulgarians-all of whom ate lamb at Easter.

It was a time before Moslems in large numbers had come from the Middle East and sheep were killed in large numbers all year round.

The veterinarian had the last word however as to whether a carcass could be passed. There were a few conditions that were acceptable to rabbinical law, but which were not acceptable to the Regulations. So that following the rabbi's examination, an inspector might well retain a carcass, only to have the veterinarian condemn it, or require that it be trimmed extensively. Or if the rabbi classed a carcass as treff (ceremonially unclean according to Hebrew dietary law), it might still be passable for non-Kosher trade. So the carcass that was Kosher had to pass two sets of standards. Rather than another ink brand like US. INSP&PASSED, the rabbi affixed a tag, with his signature and the date, according to the Hebrew calendar, and the letters, koof, shem, and resh, as read from right to left.

The posterior part of the animal was treff, only the fore quarters could be Kosher.

After the sheep were killed and before the cattle were killed, there was a short mid morning break. Inspector Jones, Dr. Christophoros and Rabbi Bekelstein were taking it easy while they did necessary paperwork.

The telephone rang and Jones picked it up, said "Hello," and then handed it over to Dr. Christophoros.

When the conversation ended and the receiver was hung up, Dr. Christophoros announced, "We are getting a trainee veterinarian, a Canadian I understand, or rather an Israeli who is a Canadian citizen at present."

This excited Rabbi Bekelstein. He had met a few Israelis, but had never had any chance to discuss things with them. He had wanted to visit the Holyland, especially after the 1967 War and the taking of Jerusalem. From his boyhood he had heard the expression, "next year in Jerusalem." Now Jerusalem was in a sense coming to him.

151

A few days later, a very tall young man came to the office, asking for Dr. Christophoros.

"I am Dr. Demitrious Christophoros," answered the doctor.

"I am Dr. Ben Levinsohn and I believe I was told to report to you here this morning for instructions. I am the veterinarian you were expecting. I am prepared to work this morning if you like."

His accent was British, rather than Canadian. He was tall, with red hair, blue eyes and freckles. He was wearing blue jeans and a plaid shirt and it would have been easy to mistake him for one of the plant employees; no coat and tie for him.

"Do you have any objection to working with hogs?" asked Dr. Christophoros.

"Not a bit of it," answered Dr. Levinsohn, "I worked with hogs at school, Ontario, that is. Oh, you mean religious objection." He smiled briefly. "No, I'm not very religious, certainly not so far as dietary law is concerned. After all, dietary law was based on prescientific experience, and the world moves on. We live and learn."

Fortunately Rabbi Bekelstein was not present to hear all that, for he surely would have disagreed. He would not have argued. He would have stated his beliefs and then been silent.

Dr. Levinsohn explained that his family had come to Israel at the end of the last century from Sweden. He had gone to school in Israel, served in the 1967 war, as an officer, and then wanting to be a veterinarian, had gone to Canada to school. After being there a year, he had sent for his wife and two children, and decided to become an American or Canadian citizen. After traveling in America and seeing the opportunities available, he had decided to become an American.

"You must understand that what you have here in America is unique." He said, "There is an incredible amount of freedom to try new things, to work for yourself and your family. Israel is a democratic state...for Jews," he added hastily, "but it is a very socialist state, like Sweden. We are in a permanent state of war. Oh, yes, we won in 1967, but the Arabs will be back.

"If I were to go back to Israel I would probably have to work for the government at a kibbutz. Sheep and goats are our principal livestock. It is too hot for pigs, even if the rabbis would let us raise them, and it is too dry for cattle."

Dr. Christophoros had never been out of the US. so he could not tell Dr. Levinsohn much about the world. He had not even traveled much in America so he just listened. He would find out that Dr. Levinsohn had a very extensive knowledge of science, and was well up on pathology, ethology, and numerous other ologies.

It seemed Dr. Levinsohn was a better biologist than inspector for there were many questions he had to ask about procedures and regulations. It was one thing to know what to condemn a hog for but something else to remember how kill sheets were to be filled out. It turned out that Inspector Jones was much better at getting paperwork correct and on time. Ben would worry about just exactly which disease the animal had and had he really been fair to condemn it, and then forget to put in the correct code number on the 403 form, or mark an X to indicate he had condemned it. But he was improving. After all, it takes years to think like a bureaucrat.

One day all four of them were in the office and Jones and Dr. Christophoros were comparing knives. Dr. Levinsohn admitted that he still didn't have the skill to get a really sharp blade. Rabbi Bekelstein let them look at his knife, but not touch it.

It had to remain ceremonially clean, not to be touched by another person. It resembled a spatula, the sort of knife with which one spreads peanut butter on bread. Except it was unbelievably sharp, on both sides. While he held the animal's head with his left hand he would make a single stroke with the blade, cutting through windpipe, and both carotid arteries and jugular veins. Unconsciousness would follow in a few seconds. He claimed that a really sharp knife could not be felt. Perhaps there was some truth to it for many an inspector has on occasion nicked himself and only felt being touched, at least for the first few seconds.

Dr. Christophoros also kept a sharp knife. It was only about six inches long and curved like a Turkish scimitar, a skinning knife. He would have let Rabbi Bekelstein handle it except that since it had been used on pork it was not touchable for the rabbi.

It remained an unanswered question among the four of them.

"Don't let it bother you," Dr. Christophoros told Dr. Levinsohn. "A sharp knife is not quite as important as a sharp mind."

Each day brought something new. Jones had served in the second world war in the Navy. Christophoros had been in the Army briefly, but had served in the Quartermaster Corp doing about what he was doing now, reviewing plants and inspecting meat. Levinsohn had been a schoolboy in Israel then. Bekelstein had tried to volunteer but was classed 4 F, having failed the physical examination. Levinsohn had served in the 1967 war on the Egyptian front.

They talked about the horrors of war and the suffering of the Jews, Poles, and Greeks.

They once talked about the concentration camps during the war. Jones remarked, "Did either of you ever know Dr. Heinz Kruger?" He was a German vet who served on the Russian front and came here after the war.

"Yes. I did," answered Dr. Christophoros. "He seemed a decent chap; never seemed to have very much to say."

"Well, he used to say that he was sure someone from the United States must have helped Hitler design Auschwitz."

"How could he have imagined that?"

"Because when he was working in Chicago in the big plants there, he used to say that when he read about the murder camps he was sure someone who was a first class packing house engineer had to have designed them. Actually, he said he didn't know anything about the camps until after the war. He said it was just like the atom bomb was here. Nobody knew what was going on and those who did said nothing about it. But looking back it was, he said, just like a big packing house for people. The way the trains brought the people in. The way the people didn't know what was going on, the use of 'kapos' like Judas goats. He had it all figured. He would say somebody from here had it all figured out and then went to Germany. I wonder who he was and where he is now?"

It was a fearsome thought and they were all relieved when the whistle summoned them back to work.

"You know", Levinsohn remarked as he filled out the reports at the end of the day, "meat inspection still bears the rabbinical stamp."

"Well, of course," answered Christophoros. "Rabbi Bekelstein would certainly say so."

"No, I know that it probably was originated by German Jewish veterinarians in the last century. But there is more than that. Look at the books you have over there. You have the Law, the Meat Inspection Act...the Torah, so to speak, with the Regulations, the Talmud, and the Bulletins and Directives...the Commentary, the Mischna. To be a

155

good inspector, one must think the way a pious Orthodox Rabbi thinks."

Dr. Christophoros swung around in his swivel chair and looking directly at Dr. Levinsohn said, "How is it that I, a fairly pious meat inspector, often fail to agree with a certain pious Orthodox rabbi?"

"That is perfectly natural. Pious Orthodox rabbis seldom agree with each other. Why should they agree with you?

"But anyway, both of you believe that any problem should have an answer which can be found in the sacred texts. And that answers can be figured out ahead of time and written down and looked up if necessary.

"Here's a humble example, which I have been told, which may not be so, but makes a fair illustration.

"A man was making a journey on foot between two towns. No one in either town knew him so he had no relatives. He had the misfortune to die suddenly on the trip, and was found by another traveler from a far country who was not a Jew. The stranger reported the death to the men of the next town. Whose duty was it to bury the stranger?"

"The town which he was closest to, I suppose," answered Dr. Christophoros.

"Ah yes, but it turned out that he was precisely midway between the two towns. What then?"

"Which town was he going toward?"

"They could not tell, he was crossways on the path."

"Then, perhaps the direction his right hand pointed," suggested the veterinarian.

Levinsohn nodded, "Yes, perhaps."

"I think I understand what you mean," said Christophoros after a long silence.

"But we don't just sit around musing about such things. People come to us and say, 'Judge between us, who is right.' How much water can a ham have pumped into it?

What is pork barbecue and what is pork with barbecue sauce? Sometimes we need the wisdom of Solomon and the patience of Job! You know, or will soon find out that lots of our regulations have nothing to do with food safety and public health, but relate to economic matters."

One day the conversation turned to religion, something the rabbi preferred not to discuss with most people, but he made an exception for the two veterinarians he worked with.

"How do I imagine God?" answered Rabbi Bekelstein. "I do not for it is forbidden in the Second Commandment. We are not to make any graven image of God. We are not to try to imagine what God is like for He is infinite and unimaginable." Dr. Christophoros would have agreed, icons to the contrary.

"Moreover, a wrong image of God will cause us to act in ways that are cruel, or wicked. Surely men who have visualized God as a cruel old king have done much evil. No correct fixed image can exist. God is far beyond our imagination."

"How do I imagine God?" answered Levinsohn. There was an outrageous twinkle in his eye as he answered. "Why, since we are said to be made in the image of God, I know that God is really a Meat Inspector, a Finaling Veterinarian. When Rabbi Bekelstein dies, God will open his stomach and find borscht and a Matzo ball. 'Kosher!' He will exclaim. When He comes to me, He will open my stomach and find a ham and cheese sandwich. 'Treff!' He will exclaim and reach for the Condemned brand."

Still, had Levinsohn really been pressed he would have admitted to being a non-believer. The rabbi had once heard Levinsohn admit as much in conversation earlier.

The rabbi had mentioned that there was not much antisemitism in the country any more. He would know, of course, even his given name meant Jew. Yet most of those he met or worked with regarded him as someone foreign and exotic. So strange and different that he could be no threat to their way of doing and believing, unlike the attitude toward secular reformed Jews who were far more numerous.

Dr. Levinsohn remarked that although he was an Israeli he had to question whether he was a Jew. Yes, his mother was Jewish. Yes, he had been circumcised under the proper ritual, but how could he claim the benefits of being a Jew if he no longer believed? If there were any. He no longer believed in God or an afterlife, or so he said.

"Look at me. Do I look Semitic, with my blue eyes and red hair? Hardly. My wife and children do because her people were from Yemen and really are Semitic.

"We are a people, perhaps. Jews come in all colors and races. Surely we are not a race. I believe Semitic and particularly the term anti-Semitic are terms we should quit using. If we mean hostile towards Judaism we should say so. Do we really need a euphemism for Judaism? A euphemism is a nicer term for something that is not considered quite nice. Sometimes we keep changing names for things as if they became soiled as we use them...like toilet paper.

"If we mean anti-Israeli we should say so. Painting our enemies with such a broad brush cannot help us reason together."

His voice had risen as he spoke as if there were great hidden anger working its way out.

Rabbi Bekelstein answered the question of afterlife, judgment and salvation when questioned by Christian friends that of course the soul went back to God who created and granted it and that surely God did not create us just to destroy us; that the next life was so different that it was totally unimaginable in present human terms, likening it to the change from fetus to mature human being.

Dr. Levinsohn remarked that for the pig, the world ended in an explosion of light, if one could assume that the pig encountered the same sensations as human victims who had suffered severe electric shock but who had survived. After that there could be no consciousness. Consciousness depended upon a biological device, the brain. When it was gone, life ended. In a sense, the Universe ends.

That morality and the civilization based upon it depends upon a belief in a God, a final judgment, and an afterlife, seemed irrelevant.

His answer was that because a myth was useful did not make it true. What is true is. Whether we like it or not does not matter. Science did not require a God. Nor did it indicate that there were ghosts or spirits, or any other sort of non-material beings. Life was not something magical, both gaseous and electrical that escaped from the body at death. Life was the result of the organization of matter in a certain complicated way as a result of blind chance.

After this, the rabbi did not have much to say to Dr. Levinsohn. He explained to Dr. Christophoros that he could not discuss or argue with someone who believed that neither God nor good nor evil really existed, that Hitler and Hillel were equal in death. This disappointed him for Levinsohn was a clever, cheerful chap with much useful knowledge. There was more. Compared to Levinsohn's gloomy view, Jews, Christians, and Moslems were in close agreement. There was a great chasm between them and Levinsohn's atheism.

He could point out that belief in God or gods, and in an afterlife was part of being human. Even Neanderthal men forty thousand years before had buried their dead with grave goods for the next world. To give this up was to become inhuman.

Levinsohn pointed out that persons who claimed to be believers had also done many inhuman acts, often in the name of religion.

One day Dr. Levinsohn noticed that a latch on a freezer door was becoming defective. He had mentioned it to both Dr. Christophoros and Mr. Jones. Jones had already complained to Sam Mordacai. But none of the three inspectors had placed a tag on the door. Being reasonable men, they had accepted Sam's explanation that the shop men had already ordered replacement parts and would promptly try to use shims to correct the problem for the time being.

There was a handle on the outside that was a lever with a heavy spring. On the inside there was a knob with a long rod running through the door that was supposed to push the lever to open the door from the inside. However if the hinge was badly worn, or the rod bent, then the necessary connection would not be made and the rod would not push the lever and the door could not be opened from the inside.

Time passed and the door latch was forgotten until the handle finally sagged enough that the rod ran over the lever rather than into it. It could push only empty air. Jones had meant to tag it, but had not gotten around to it. Somehow, when he had been ready to tag it he had run out of US. Rejected tags and on the way back to the office he had been interrupted by other duties and problems. The maintenance man had again promised to fix it anyway.

They had all forgotten it when they had been called into the freezer to look at a glacier of ice which contained some boxed beef. Somehow the freezer had an air leak that allowed outside air to get into the freezer. Water in the air freezes on cold surfaces. In time ice builds up in large commercial freezers just as it does in the freezer compartment of a household refrigerator. Then the whole thing must be defrosted. If the freezer has a bad roof into which rain can seep, the process runs even faster. The door was slightly open when someone outside pushed it shut. It was the normal thing to do to save electricity.

They looked at each other. There were three of them: Dr. Levinsohn, Sam Mordacai, the owner, and Dr. Christophoros. Jones had stayed in the office to fill out papers. Had he been with them, he would have had the presence of mind to block the door to prevent its closure while they were inside. When Jones finished he went home, ignorant of the plight of the others. Each in his own way was responsible for their dangerous predicament. Any one of them might have noticed the condition of the latch and blocked the door with a box, or could have done more to have had the latch repaired. The temperature was minus 20, and there was wind. Dr. Christophoros and Dr. Levinsohn while in the freezer had heavy coats that they had taken from the hanger just outside the freezer. Mr. Mordacai had on complete freezer clothes and could be fairly comfortable.

Had they not grabbed the coats, their condition would have been far more serious. After they fought the latch for a few moments and gave up, Sam grabbed a box. Each in turn set a box on top of another. Sam was able to get up to the top of the pile of boxes and cut off the power to the fans. Fortunately the freezer had a fairly low ceiling, only fifteen feet. Some freezers go as high as thirty feet or more today. Then it would not have been possible to get up into the machinery and cut off the freezer fans. The freezing wind stopped. It was still very cold. Soon their hands and feet would freeze. They piled the boxes up next to the door and pushed open the flapper that fit around the rail that entered the room. Then they shouted for a few minutes but no one came to let them out. There was too much ambient noise, and most of the plant workers had gone home.

At least the opening would let air into the room so that they could not suffocate. There was an irony to this because it was unwanted air leaking into the room that had caused them to be there in the first place.

Had the room been built as a freezer from the very beginning there would have been no opening for the rail to run into the room, which was a major source of moisture laden air leaking into the converted freezer.

Next they built a shelter out of boxes, as best they could, and got inside. It wasn't possible to really seal it in because there was nothing to cover the top. Then they remembered that Dr. Levinsohn was still carrying his scabbard with knife and hook. Sam took it and cut open two or three boxes and put them on top. The box had contained internal fat from cattle. He had piled it all into another open box so it would not be contaminated and might be repacked.

They realized that it was fuel that could be used to keep warm if they could find a way to burn it. That required a wick, and matches to light it. Dr. Christophoros found a cigarette lighter in his pocket and a handkerchief. Using more cardboard they rigged up a candle inside their little shelter. They wondered about danger from carbon monoxide, but thought that the opening would allow enough circulation of air to be safe. At least they could warm their hands.

They sat down, each on a box, and pondered their plight. It had been only ten or fifteen minutes since they had first entered the freezer. It had seemed much longer. They thought about how long it would be before they were missed.

No one would come looking for Dr. Levinsohn. He lived alone and had irregular habits, especially since he worked all the available over time he could. Sam Mordacai looked at his watch. It said 5:45. Everyone had gone home. Surely they would notice that his car was still there. Sure, he thought. It often stayed at the plant when he went home with his brother, or son in law. Dr. Christophoros usually took the bus or streetcar, having showered in the inspection office. He would not be expected home until after seven. Sometimes he missed his bus and was late. His wife would conclude he was simply late and would not become alarmed until it was much later. Then she would call one of the inspectors to find out when she could expect him. She really would not begin to raise any help for some time. It looked as they would be keeping each other company for a while.

Sam said, "Anybody bring any cards?"

"Nope," answered the other two.

"Too bad the Rabbi isn't here. We could discuss religion," remarked Sam.

"Like the story of Jonah and the whale," answered Dr. Christophoros. He had thought of the three days in the tomb following the crucifixion, but thought to spare his two Jewish companions.

Levinsohn laughed. "We might be here for three days at that."

"We better not," said Sam. "There will be a night watchman fired if we don't get out pretty soon. Come to think of it, we ought to have some hot dogs to be temperatured and taken out around midnight. That means you have an inspector due in the building as well. Well, at least we have a chance to talk. It's not likely that we will be interrupted any time soon. I'm not sure we would be safe sleeping, between the cold and the carbon monoxide from our little fire here. How are your hands and feet, Doc?"

"O.K. now, but they sure were cold a few minutes ago."

Addressing Dr. Christophoros, Sam said, "We damn near got killed. I'm sorry I got you guys into this mess. I know we talked about fixing the door, and we would have gotten around to fixing it next week, but we didn't have the part and that takes time."

163

"We know, that's why we didn't insist on tagging it up right away."

Time passed, they talked.

Dr. Christophoros remarked to Levinsohn, "You were in the Seven Days war, weren't you?"

"Yes, I was an infantryman. I was a lieutenant."

"Funny," said Sam, "I was an infantryman in the Pacific in World War two. I wound up a sergeant. Saw lots of action. Sometimes I still dream about it. Mostly I dream about the jungle valley. We had a jungle valley to cross, a trail to follow, and there were only a few of us. We were certain the Japanese would ambush us on the way. Skeleton death grinning horribly was waiting to snatch me from off the trail.

"Death was hiding behind one of the trees I must pass somewhere in that dark dense jungle. Far away we could catch glimpses of the sunlit mountain peaks we were trying to reach, but we knew we would never get there. Death would snatch us first, one by one.

Sam paused, "We were lucky though. The Japs had pulled out and we made it through okay."

There was another worry that Sam decided not to share with his companions. Shutting off the cooling fan might cause a buildup of pressure in the ammonia lines. Should the line leak or burst inside their room the ammonia would kill them.

He had a moment of black humor. Ammonia is an explosive gas. There would be an explosion if there were to be enough ammonia along with their little fire. It would probably blow the freezer door open. Being dead and free would be a poor bargain.

"You know, it's Yom Kippur tonight," remarked Sam. "The shofar (the ram's horn announcing twilight and the beginning of Yom Kippur) has surely blown by now. If we were Observant we wouldn't be here in this freezer."

"If we were observant, we would have tagged the freezer." retorted Dr. Christophoros.

They could not have known it but the war had begun again in the Middle East. Egyptian soldiers were crossing the Suez canal and attacking the Bar Lev line. Pretty soon, Sam looked at his watch. They had been in the freezer for four hours. The watchman should be in the area, he thought. He climbed up to the top of the door again and pushed open the rubber flapper around the rail. He heard a distant whistle. The watchman was whistling "When Irish Eyes are Smiling".

Sam let out a yell, and in a few moments the watchman freed them. Sam put out the fire and turned the fans back on. They looked at the boxes in disarray and then at Dr. Christophoros and said, "Lets leave 'em til Monday." And they did. "Lets go home!"

Monday came. Jones tagged the freezer. Then the freezer was defrosted. A few days passed. The door was forgotten temporarily until the repairman called the inspectors to check on the repairs to the door and to remove the tag Jones had put on the door.

They looked at the door of the freezer, the one which had almost killed them. The latch had been repaired.

But something else had been added. There was now a Hebrew inscription on the door jamb.

"What does it say?" asked Dr. Christophoros.

"May the Lord bless your going in and coming out. May the Lord bless and keep you," answered Dr. Levinsohn. "An appropriate mezuzah, I'd say."

"I'm Dr. Wadloe," said the young man who came into the office the morning of the next day.

"Dr. Levinsohn...wasn't he supposed to be here today?" asked Dr. Christophoros.

"I was told he had resigned and returned to fight in the Irgun."

"As suddenly as that?" asked Rabbi Bekelstein.

"I suppose so. They told me very little."

"I will pray for him," declared Rabbi Bekelstein.

"So will I," said Dr. Christophoros.

" 'You will return,' saith the Lord, 'not because you have willed it but because I Jehovah have willed it.' " remarked Dr. Christophoros, but he could not remember chapter and verse, or even which book it was from. He would have to go back and read his bible again.

Regulatory Wheelies

"Did you hear about Dr. Smith?" asked Jose Ricaro, the inspector.

"Hell, I was there when it happened!" answered Williams, the other inspector.

"Is he going to be okay.? Sure, he's a mean old man and all that but I sort of like him. He always was square with me."

"Yep, me too. They say so. Seems he had a mild stroke. So far he shows no paralysis, though. First they thought it was a heart attack. I suppose he'll take medical retirement, he must be nearly seventy. He started in '33 and here it is some forty years later."

"Well, since you were there, tell me what happened."

"Okay, Jose, it was like this. It had been a typical slow easy day at Pintafour Packing, no problems much. We killed forty nice steers. Doc was on heads and finals. I worked guts and rail. We were in the office writing up the

kill reports when the phone rings. Doc picks up the phone, put it to his ear, listens a few moments, fiddles with his hearing aid, puts the phone back down on the receiver cradle and mutters to himself, 'Silliest damn dream yet,'

"The phone rings again, like it was urgent. Doc reaches over and bangs on the phone like you bang on the top of an alarm clock to shut off the alarm. After he bangs it a couple of times he looks up with a glazed look on his eyes, leans back in his chair and passes out.

"Mind you the phone is still ringing. I pick it up. Fuller from Washington is on the line. 'We've got real trouble,' Fuller continues. 'Yes,' I answer. 'Dr. Smith just passed out and we need to get him help. How did you know? Please, can you call us back in a few minutes, or let me call you!' And I hung up, and rang the main front plant office and the rescue squad came in about ten minutes. They took Smith to the hospital. Just as they carried Doc away in the ambulance the phone rings again. It's Fuller again, mad as a hornet.

" 'You guys cut me off and I've got a lot more plants to call. I can't raise your Regional Supervisor. Where the H is Dr. Delancy anyhow? Yes, I know it's Friday afternoon. Have you any idea what time it is here in Washington, D. C?

" 'The ink, the purple ink we have used for damn near seventy years for the inspection brand CAUSES CANCER! It is supposed to have caused bladder cancer in rats in Germany. You people have got to cut off all the inspected and passed brands with the purple ink and use some other kind of ink.' "

Jose laughed. "You're kidding me! Really. Did Fuller tell you that!"

"No bull, it really happened," declared Williams.

"I wonder what the rabbi would have said?"

"Thank God we don't have a Kosher kill here. Anyway I was still on the phone with Fuller, so I say, 'Well, what would you suggest? I mean, what color? I've got to tell these plant people something, they might think I'm

168

wrong. This is a bit unexpected, you know. I better call Mr. Halley, the plant manager, in here and let you tell him too.' (I wanted to ask him if he was really our Dr. C. J. Fuller in Washington and not some guy pulling a joke, but I had gotten to know Fuller and it was him all right.) About this time Mr. Halley walks in and asks about Dr. Smith. I stopped for a moment and said, 'Mr. Halley just came in; you can tell him too.' Then I hand Bill Halley the phone. He knows Fuller too.

"First Halley's face is white, then it's red. He puts his hand over the mouthpiece and asks me, 'Is this joker really Fuller?' In the meantime I can barely hear angry noises coming out of the earpiece, 'I didn't write that infernal Delaney Amendment and I'm not Mrs. Foreman! And I'm telling you that you have got to use some other approved ink for now. Call your supplier. Oh, any color except green, that's for horse meat. Be sure it's an approved dye though. How about some hot dog dye? Got any on hand?'

" 'No, we don't stain our hot dogs out here,' answers Halley. 'You remember that. That's a Southeastern custom. Besides we don't make hot dogs here. Sorry. Tell you what. Can't we just seal the cooler until Monday and maybe we can go from there?'

" 'Okay, I have more people to contact. God knows what the public will think. Good luck. Let me talk to Mr., ah, the Inspector , Mr. Williams again.'

"Mr. Halley waited a moment until I hung up and asked ' Does your wife leave the brands on the meat she brings home? Mine won't, even though it's supposed to be harmless.'

" 'Mine too,' I said. We will all hear about this one. Anyway we have only the two kill floor Number 1 brands to remove. None of that climbing around until we rebrand.

" 'Maybe common sense will prevail in Washington by Monday,' said Halley.

'Don't bet on it,' I replied," Williams answered.

The major large pieces of meat each require a brand. They are called primal cuts. However only one brand is required for the animal to be passed off the kill floor. One is required for each piece and since carcasses are split down the middle only two marks are used. Both the embossed metal marker and the mark it makes are called brands.

Why? Because the cattle carcass is usually covered with a wet sheet of muslin, wet with salt water having the same salinity as the body fluids. When the carcass it taken out of the cooler it is dry and the brands, or marks of inspection, are much more legible when placed on a dry rather than wet carcass. It is usually then that all the parts are branded.

"So that's why the cooler is sealed and the carcasses are still in there. I think we are going to bone out the whole bunch and grind the meat rather than send them out with some strange color the customers don't know and trust," concluded Williams.

At that moment, in walked Dr. Karl B. Delancy who said, "Either you guys know the latest about Doc Smith? I was able to visit him Sunday. His wife wants him to retire.

"So do I. Or at least I'm not sure I want him back 'til we are sure he's well. I'll be doing finals here for a while. Hasn't the kill started yet?"

"I don't think we are going to have a kill here today. Mr. Halley and the front office said they would probably decide to bone and grind the whole bunch today," answered Jose.

"Hm, odd, usually the stuff here is too nice for that." Dr. Delancey hesitated a moment as he put on his scabbard, snapped the chain, and then put on his helmet.

"Anyway about Dr. Smith," continued Dr. Delancey, "I asked him what he could remember about his passing out. Said he knew he worked okay Thursday. Thinks he must have had his stroke Thursday night. He had the funniest dream just as the alarm clock went off. Said he dreamed that Fuller called him from Washington after the Friday kill to tell him the purple ink we have been using for all these years caused cancer in rats and he was going to have to cut off all the brands and use chartreuse or some other color ink, anything but green. Said he couldn't wake up and couldn't shut off the clock, and then suddenly he was there in the hospital Saturday afternoon. Wasn't he here through the kill Friday? And if not, who took his place on heads and finals? Of course, I suppose Jose could have done processing and helped out. Of course there is usually nothing retained here. But I still thought he worked Friday. He must have had the stroke Saturday morning and just couldn't remember Friday."

He took his helmet and scabbard off again and sat at the desk and filled out some papers and opened an envelope with the latest manual changes.

"Anyway I had a good trip to Montana and saw my folks. You ought to try it. You can't beat the solitude for improving your outlook.

"Got to do a lot of shooting, out there in those wide open spaces. There's nobody out there to object.

"That sure must have been a weird dream old Dr. Smith had. He sure has some strange ones doesn't he?"

Ed Williams and Jose Ricaro looked at each other but neither could say anything. Perhaps they were afraid of causing another stroke.

Eventually new dyes were used. It was hoped that these would prove safe to the nth decimal point. Cynics remarked that virtually everything caused cancer. Cooking, especially charring the surface of the meat was hazardous.

Dr. Smith recovered from his stroke but decided that retirement was the best promotion he could have. He might have gotten a promotion to a desk job but didn't want to be

171

stuck behind a desk. He dreamed of green pastures out with the hogs under the bright Illinois sun. He could see his brother in Texas, as well as go out to the Hershfeld farm where Timothy lived and farmed with his wife and children. In time, the farm would be real estate. It was not too far from the great accelerator laboratory the government had built.

He and Elsie went out to Texas to see the old home place. Not much had changed. His brother had built a new house near the old one, but the same fields, creeks, and gullies he remembered as a boy were there. He could take a .22 and pop off at jackrabbits. Before he could completely go to seed, Elsie took him back north to the Hershfeld farm. Timothy had gone to Iowa State with the idea of being a farmer like his grandfather. However the farm was being swallowed up in a sprawling suburbia. To remain a farmer he would have to go west, which he did soon after his parents' visit.

At this point an explanation of the nature of bureaucracy is due the reader.

If one works for a large organization, one is a tiny brick in a huge pyramid. Actual work takes place at the very bottom where the bricks rest on the ground. As the pyramid becomes larger the volume increases in relation to the area covered at the base. The inspectors I am writing about are at the bottom where the real work takes place and where the pressure due to the weight of all those above presses down upon them. Also as the pyramid grows larger those inside are further removed from the base, as well as from the outer faces where the winds of public opinion can be felt. From top to bottom messages are passed up and down and sometimes are lost or diluted.

As with any industrial job, there are dangers in working in the meat industry. Some jobs are more dangerous than others. In a packing plant there will be electricity (up to 440 volts) and boiling water, some of it at high pressure. There are dangerous gasses; ammonia, freon, carbon dioxide, liquid nitrogen, natural gas, and propane, as well as compressed air and steam.

These are dangerous if mishandled. There are bright lights from arc welders, too bright to look at. The noise can be actually deafening, well over 85 decibels. Finally there is a host of carnivorous machines: machines built to cut, slice, grind and cook meat...of which we are made.

Yet most accidents are minor. Simple falls lead the list as a cause of injury. Collisions with objects like tables are second. Grease and water mean slippery floors. Boots with special treads, handrails, and perforated steel grates to stand on all serve to reduce the danger.

Minor knife cuts and burns are next. Knives, saws, and other equipment need to be kept clean, in fact sterilized, which is done with very hot water or steam.

Dangerous as it is, it is probably as safe as working at home. Inspectors are supposed to avoid risk, and to set an

173

example to others in obeying the rules of safety. However they are not held responsible for safety other than their own.

An inspector involved in an accident with resulting injury must file an accident report even though the injury seems trivial. This is both in the interest of future safety and against the possibility that the injury is more serious than first appearances suggest. An example could be an infection from a minor cut, or a fall resulting in a back injury. Almost anything one can run into will be hard and unyielding, as well as perhaps sharp, even a fellow worker or inspector.

A modest example follows.

"Ouch! D----n it all!"

"What happened?"

"Hey Doc, I just cut my finger."

"Bad?"

"Naw, just a lil' nick."

"How come you done that, ain't you workin' rail?"

"Just cutting myself a little chew."

A brief explanation of the above is as follows. Veterinarians and other inspectors are all called "Doc." That is, the packing plant workers call all the inspectors "Doc", and the lay inspectors call the veterinarians "Doc". It is an honorable, if sometimes honorary title.

In due time, the inspector or his superior will file a report with higher authorities showing that Inspector X cut himself while sharpening his knife preparatory to relieving Inspector Y who was incising mandibular lymph nodes at the "head" station. Smoking in the food and product areas is forbidden, so is expectorating on the floor Obviously he did not have a knife at the rail station since the Manual of Procedures forbids it and explaining this breech of procedures would lead to still more reports to be written and filed. The report will recommend greater care in knife handling.

At heart a meat inspector is a bureaucrat. So am I. In the Greek chorus of life our livery is brown. Politicians propose but we bureaucrats dispose. Politicians declare what should be done but seldom stick around long enough to become involved with the real work of carrying out the declared objectives. They move on to other things. We stay put. Procrastination and obfuscation are not vices but essential tools. The humble bureaucrat knows that the decrees and directives that cross his desk require a sort of triage. Of matters that affect him or his department, he (or she) must decide whether an item is useful, desirable and within his ability to accomplish. Is it desirable, but not practical? Or is it neither desirable nor practical?

Suppose the directive requires that a hole be dug. Perhaps he has the equipment for digging holes, likes digging holes, thinks that a supervisor whom he intensely dislikes will surely fall into the hole; the hole will be promptly dug. It will be wider and deeper than the directive actually required.

Or perhaps he agrees that the hole is needed but has none of the hole digging machinery. He has wanted this equipment for some time but has been unable to persuade his superiors of the need for it. So he sends in his requests again pointing out the directive as his reason. Excavation will have to wait pending the arrival of the machinery.

Or suppose he knows holes are a foolish idea, dislikes holes (having fallen in on previous occasions) and knowing that having been dug, they are invariably in the wrong location. Then they have to be filled up again. So the end result is the status quo ante. Moreover, soon new policies and directives will follow the election of new officials.

Delay and obfuscation will save digging and filling an unnecessary hole. Soon holes will be out of fashion. Pooh Bah explained this to the Mikado long ago, in his description of the execution of Nankipoo: how it was "as good as done even if it were not done."

There is almost an infinity of terrible things which might happen but probably will not. That is; they seem very unlikely. However, should the imagined calamity occur, then the bureaucrat will be held responsible. For example, every few thousand years a large meteor strikes the earth. Should it happen now, NASA and the DOD will be asked why they permitted it to happen. Or if you are the Mayor or Chief of Police in a very small town and suddenly thousands of people decide to visit in the absence of a law against it, you or your council will be accused of lack of foresight.

During the civil violence accompanying the war in Southeast Asia, we feared terrorists would poison the food supply, and/or introduce exotic diseases. There are poisons that would have escaped our gross inspection techniques then available.

There was an accidental addition of a fire retardant chemical to cattle feed in the northern Midwest. Hundreds of cattle were poisoned and had to be killed.

Pork from hogs fed grain which had been treated with a mercury fungicide poisoned a few people in the Southwest. This grain was intended only for planting (not feed) and was so labeled. Some how they failed to understand the warning or chose to ignore it. The poison was fed a little bit mixed with other feed each day so the hogs survived it, but when the hogs were killed and eaten, the people who ate the pork received a much larger dose at one time.

We now have a sampling program to help prevent such occurrences that is improving with time and experience.

The story about Dr. Smith and his stroke is fiction but the part about the purple ink is absolutely true!

One day we received a call to stop using the purple ink we had used for over half a century to brand carcasses

"US INSPECTED & PASSED" because the ink was linked to bladder cancer in rats according to an experiment in Germany. It is now thought that internal parasites were the cause of the bladder tumors and that the experiment was faulty. Life is stranger and funnier than musical comedy.

Often the public is best served by bureaucratic inaction as Admiral Rickover pointed out. Or to quote Milton, "They also serve who only stand and wait." However, not being engaged in vigorous activity sits badly with the working tax paying public. How does one cope with this?

Besides keeping a low profile, it is sometimes useful to enforce more law than the public really wants enforced. Raid the local church bingo game to enforce the gambling laws. Be sure to embarrass an elected official who has been calling for more effort from the public servants than seems reasonable.

Another problem is that if the bureau is seen as underemployed or underutilized, then the bureau will be given additional tasks. The bureau may even welcome them, but the problem is that these new duties may come to be seen as the main function of the department. When the firemen are called to put out a fire, they may be out delivering toys.

It is often alleged that most regulatory agencies are in league with the industry being regulated. This is true, but is not contrary to the public interest. Meat Inspection serves as a board of ethics for the meat industry. Inspection cannot function effectively in a manner contrary to the interest of the industry. It enforces rules the industry believes are good for the industry and the public. It works this way.

Companies A and B are industry leaders. Well run, efficient and profitable, and serve the public well with good sound products. Those products cost more than the products produced by companies C and D who are less efficient and cut corners on quality. C and D may want to improve, but cite competition by F, a company with very low standards. Eventually A and B will have to lower their standards if F can undersell them. A, B, C, and D believe this to be bad for the industry and the public. The political support for the inspectors who force higher standards on company F must come from A, B, C, and D rather than from the general public. Inspection cannot enforce a totally uniform standard. However, with the help of the industry, over the years the record has been quite good.

Since the inspection service relies to a great degree upon the support of the industry, it could not function if the industry were hostile. Company F will be forced to raise its standards or leave the business, but the real pressure to do this will not come from consumer advocates, the public or the government, but from its competitors who know that public confidence is necessary for a prosperous industry.

There is another facet of regulation that is important. It is not possible to measure and enforce at the same time. An example is that one cannot use a micrometer as a C-clamp and expect it to maintain its accuracy. It will stretch and give larger readings as it stretches. The same sort of effect takes place in human affairs. Both individuals and organizations have great difficulty in enforcing standards to the exact point really intended by those who originally set the standard.

There is a very human tendency to soften the standard to fit the actual case. There is also the tendency to get provoked and enforce too high a standard. What is often needed is someone to provide the force required and someone else to say whether the objective has been reached... or exceeded. That is why the police are separate from the courts.

Regulatory organizations generally have others watching them to determine whether they are performing properly, and to insure uniformity and fairness throughout the system. In meat inspection this has been called compliance and evaluation. It serves as the micrometer while the inspection service provides the C-clamp.

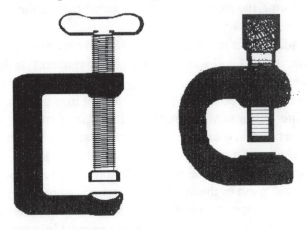

Meat inspectors have two general duties they take very seriously. The first is to assure that nothing that can harm the public gets into the trade channels. The second is to protect the livestock industry. After all, they are sampling the animals produced by our farms and ranches. They are often the first to detect disease.

When they do, they report their findings to the field service, another branch of the government, whose men go out and attempt to trace back the source of the affected animal. When they find it they may issue quarantines, and then depending upon what they find, may treat the animals, or require that they be sent directly to slaughter, with shipping papers, or if the disease is especially dangerous, they may require that the animals be killed and buried on the spot.

As for the public safety, most of us go by the rule, "If you wouldn't eat it, then don't pass it." A good and practical rule of thumb, or stomach if you prefer, often used to cut regulatory Gordian knots.

Way Down South

About the time Smith retired, another young man was also leaving the meat industry in another part of the country.

Randal Sharp had had enough. He would take no more abuse, neglect and scolding from his two uncles, or his father either, for that matter.

He thought about the years he had put into the family meat packing business his grandfather had started in the years between the first and second World Wars. H. H. Sharp had originally been a pig farmer on the Southern coastal plain of North Carolina. He would take a hog, kill and dress it, and take the meat to the small rural grocery stores to sell. You could say it started first with one hog, then two, then a dozen, and after two decades and the war years it became nearly a thousand a day.

Old H. H. had died when Randal was a baby, but he heard many tales about him. People in the family said he was too trusting and easy going. Yet outsiders all spoke well of him. During the depression, he let folks run up debts that they sometimes never paid. However there were others who could and did pay or lend him enough money during the war to greatly expand his business.

Randal's father, Preston, and his two uncles ran the business with a tight grip. Uncle Horatio was the eldest. Uncle Beauregard was the youngest. Some time later after Randal had left, they sort of ganged up on Preston to force him out of the business. Despite being partially crippled by infantile paralysis as a child, Beauregard Sharp ran the kill floor and the processing of pork.

Beauregard, or Uncle Bo, as the help called him behind his back, was a formidable character. He used a cane to help himself stalk about the plant. He also used the cane to emphasize points of argument. Legend said he had laid out an assailant in the back streets of Raleigh on more than one occasion.

Uncle Horatio stayed in the front office and the loading dock and sold the meat. His son also helped.

Preston bought hogs, weighed them and sent them up the lane to the stunning pen. He was largely guided by the principle that the cheaper he could buy a hog, the more profit there would be.

Not far to the East was a good and growing source of hogs, a place called Pantego. There was, however the presence of larger more powerful folks in the same meat business not far away. If Sharp Brothers were too successful, those other folks might become jealous and a price war could break out. It would be one they could not win but which they all might lose. So there were limits they felt they could not cross without trespassing. There was never anything ever written, or hardly ever said but it was there.

Just exactly what caused Randal to decide that this particular straw would be the last one was not certain.

He was standing outside the plant at 5 AM. He had just gotten off from school for the Christmas holiday and driven home late the previous night. He had slept three hours and the gotten up to help work the kill floor. The day before he had finished a particularly mean exam on a course required for his MBA. If he succeeded he would get his degree the next spring. Having finished his engineering degree at VMI, he had a commission awaiting him in the military but he had managed to delay, to get an MBA. He had first studied mechanical engineering and then changed to business. His father had thought that an education would help. School seemed a happier, more rewarding place than the old packing plant.

Neither his father nor his uncles had been to college. They had been victims of the great depression. It was sad,

Randal would reflect in later years. College might have done no more than widen their very narrow view of the world and business, but that would have been enough.

Randal thought back about how he had driven hogs for his daddy and uncles. Then he was too small to even see over the lane and its parallel fences which led from the holding pens to the shackling pen. They had taught him to be mean, at least to hogs. Later he would unlearn it, but then he was grown.

Uncle Bo would play shuffle with the workers' time cards, so as to keep them working even when they wanted to go home. Some said he was not above cheating them on their time, not so much from avarice, but simple meanness. His workers managed to cheat him in turn. They often did things the wrong way accidentally on purpose. They would pretend to be furiously busy while Uncle Bo was watching but fell slack when there was no one watching.

Every so often, Uncle Beauregard and Uncle Horatio would hire someone with real packing house experience to work for them. The idea was that having a real foreman would allow Uncle Bo time to think and do more important things. There were several competent, experienced foremen who came to work, but left after a year or so. This was because Uncle Bo would not let them do their job. They simply were not permitted to reward or punish those working under them. Moreover, they weren't allowed to function as management. Invariably, a worker would be missing and the foreman would be pressed into working the line somewhere. Then Uncle Bo would play foreman, often undoing what the other man had spent time and effort to get someone else to learn to do.

If you had spent time learning a difficult job like eviscerating a hog, and expected to be paid extra for doing a better than average job, you would expect to be allowed to keep on working at it and making the extra money and

182

becoming even more expert at it. But just about the time it seemed your efforts were ready to pay off, you were moved back to some other lower paying job. Whether you were an inspector or a plant employee, you couldn't go to the foreman for answers or instructions. He too had been demoted and was simply working the line.

Randal thought back to the innumerable times he had brought something to the attention of his father, or his uncles, and been told that it was a silly idea. It had come to a head because Randal believed in working smarter rather than harder.

There was more. Unless they would work smarter, the business would go under. The competition was going to keep on raising the level of required efficiency and money would be flowing into the business, only to be wasted, rather than out of it to advance the business.

How many examples could he think of? Should he start from the beginning, or from worst example...well maybe a combination. Not all of the good ideas were his, but he or someone had presented them and they had been rejected.

One of the foremen had suggested that the hog carcasses should be graded. Whole carcasses that had no defects nor blemishes should be separated and sold, for a premium, as whole carcasses to the retail trade. Carcasses that had cuts, blemishes, or bruises should be sent to their own cut floor and the meat cut up. These too should be graded with the worst parts going for boning for sausage. This would however require training an employee for this job, as well as an understudy, who could do the job in the employee's absence, and letting the employee do that job without interference. Uncle Bo simply couldn't bear to leave an employee alone to do the assigned job and then judge the results. Especially if the job involved thoughtful observation and decision rather than gross and furious physical activity..."real work".

There was something else. Randal tried to get his father to keep accurate records on how hogs from various farmers dressed out, to say nothing of keeping up with who sent in hogs that the veterinarian later condemned, so as to avoid buying bad hogs. He remembered one time the plant had been filled with "possums" (so nicknamed by the employees) and the line had been completely jammed up with stunted, arthritic, pigs barely beyond feeder pig size.

They had to stop killing pigs and put everyone to trimming those pigs already killed. Eventually most of them were condemned. They had been cut to pieces in an effort to trim them. Many were so emaciated that the veterinarian condemned them with scarcely any examination. It may be that in the end they lost count of how many were U. S. Condemned and how many were simply cut down by plant employees and tanked. It seemed that Uncle Preston had bought a bunch of pigs cheap, real cheap, and had been misinformed as to how old they were.

It made a very long day. His two uncles had come in and finally had the remaining piglets put into the tank as if condemned. Many had been condemned. Most of these were joking referred to as U.S. Number 7, 8, or 9.

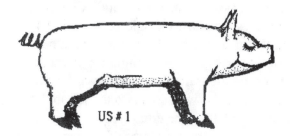
US # 1

Actually there are only three grades of hogs, U.S. 1, 2, or 3. and below grade.

It took the veterinarian and the inspectors a couple of extra hours after the kill to get all the papers filled out.

US # 2

Most of those little pigs called possums were victims of a very severe infant diarrhea. They had suffered an eversion of the anus and rectum. That is, the end of the intestinal tract had turned inside out. If you had seen them then you would have wondered why they had red sausages hanging out of their behinds like extra tails. Then soon the everted gut would die for lack of blood supply and be sloughed off leaving a very much smaller orifice for the passage of feces. So they could eat for a while but were very constipated. Losing your anus is no joke, but a very serious matter for man or beast. Even had the piglets been repaired surgically they would have had no bowel control.

US #9

Prevention was and is the only sound treatment. Gradually, Uncle Preston was persuaded of the advantages of buying better hogs. It became obvious that the final price of cheap hogs was expensive.

185

Originally, the Germans had developed a process of finding the largest artery on a ham and then adding a precise amount of solution to the ham. This was a tedious labor intensive process, but it was very exact when the ham lay on a scale and the dial showed what the end amount of added weight should be. Once the water with its added salts and spices was in, the artery was tied off. Then the ham was cooked in a smokehouse. The solution would not run out, at least not very much.

Later, a device called a stitch pump was invented that had numerous large hypodermic needles, which were forced into the ham, and then the solution was injected. The ham was run through the machine twice, once for each side. The amount of pressure determined how much fluid was added to the meat. Once added the fluid would run back out so the process wasn't very exact. You might say, as many packers did, that at least water is a wholesome, non-fattening ingredient...cheap too.

That was the problem. Ham is much more expensive than water. Inspectors would weigh hams before and after. Sometimes the same hams, sometimes other hams of the same weight range. The idea was to test the procedure rather than just the product. If the procedure was correct, the products would be right.

When they had first gotten the stitch pump, it had been set to too high a pressure. The first ham had its weight more than doubled. Uncle Horatio and Uncle Bo looked at the ham, bloated and flaccid like a beached jellyfish and about as appetizing. Uncle Horatio said, "Lordy, if only we had had one of these machines before we got inspection!" They all laughed and then threw away the ham.

Randal had just finished a course in statistics which he immediately applied to ham weights. He showed the processing foreman just what the odds were of getting in trouble with the inspection regulations based on how closely the amount of water added could be regulated. Something he called a standard deviation; a number based on how much the hams varied in water content following pumping and cooking. The tighter the water variation could be controlled, the more water could be safely added. He called it walking near the edge of a cliff without falling off. The less you stagger and wobble, the closer you can safely stay near the edge and not fall off.

Putting the maximum allowed added water into each ham required close control so as not to put too much water into any one ham. The Inspection service had penalties for over pumped hams that were no fun for either inspectors nor packers.

But controlling the water better meant having more modern stitch pumps and taking far better care of them than had been done in the past. The result was that to stay out of

187

trouble, most of the water had to be cooked out of the hams, and on any large number of hams meat had to be given away to the customer to stay out of trouble with the Inspection Service. Perhaps as much as five percent of the net weight had to given up because of inaccurate pumping. That is a lot in a business that often only makes a few percent on a sales dollar.

Randal's uncles simply would not believe that such a thing was real. Even more surprising was a stepwise progression in a series of cooked ham samples that showed according to a goodness of fit ratio that something was happening to the hams that took about six percent water out of some, about twelve percent out of others, and even about eighteen percent out of a very few hams. These were hams which had the bones removed and then were placed in molds and cooked. Then there was nothing but a block of delicious ham meat ready to be thin sliced and put on biscuits or sandwiches. They were very expensive, both for the packer and the customer.

There were three of them looking at the operation; the inspector, the processing foreman, and Randal Sharp who had stumbled across the phenomenon. They wondered if some had been cooked longer, perhaps overnight, or whether the temperature had been high sometimes, and so forth. Suddenly the answer appeared. The boned out ham meat was put into a mold which had a spring loaded top held down with a ratchet. An elderly black man would load the mold, place it onto a rack, and then pull a lever compressing the springs that squeezed the top into the mold. Sometimes the old man would have to pull the lever twice, very rarely three times.

As this went on, Randal watched with mounting excitement. He held his hand out pointing and almost quivering.

The foreman understood immediately, "So there's the guy putting the squeeze on profits. Each squeeze is worth six per cent water." He was quiet for a few moments. "And you knew something like this was going on just from your test results and your slide rule."

Actually it had been more complicated. He had used a mechanical calculator with wheels and registers. It was just before electronic calculators were available.

Unfortunately, when he tried to explain to his uncles and his father what he had discovered, they paid no attention. What could a young pup like him with his fancy math and book learning tell them about a practical business like theirs?

The answer was that had they listened he might have saved the business. It was the beginning of America's contest with Japan in industry. Dr. Deming's messages concerning quality control had not yet been preached throughout the land. Quality must be designed into a product from the beginning. It was not practical to try to inspect quality into a product.

There had been an even earlier time when Randal had first worked in the pens, and then had been called into the plant to do the trimming at the rail station. He had observed a plate in the floor in the holding pen upon which hogs slipped and fell. It was made of steel and was very slippery when wet. When a hog's feet slid out from under him each foot went to the side, putting great stress on the adductor muscles. Hogs were not designed to do the split, so when the hog got up having had its feet slide out sideways, it could then hardly walk. The hams would have large bloody bruises or tear in the meat and the hams had to be trimmed and downgraded.

He suggested that lots of bruises and injuries could be avoided by following up each injury from the trim rail

189

and finding out where it happened and then changing or removing the cause.

He was met with derision by his father.

"This is my part of the plant. I won't have them bothering me down here in the pens. They don't care so why should I? They won't agree to spend what it takes to fix up the pens better, so let them worry about it up there on the kill floor or the cut floor. Probably the saving would be so small it's just better to crank faster and not look back."

He meant, "Work harder, not smarter," although Randal had not understood it that way at the time. Neither of them gave any thought as to whether it made any difference to the hogs. They were soulless objects, and it would be many generations before that attitude would change.

The black men who worked at the plant and who usually drove the pigs would sometimes show some sympathy. Every so often, a hog would escape from the plant or the surrounding pens. Usually this was when pigs were driven off of the trucks into the weighing pen.

Outside, the plant and pens were surrounded by farm fields on one side and deep woods on the other. A pig might be able to subsist in the woods as a wild animal for a long time if he had street, oops, woods smarts.

If the pig escaped during a work break for the workers inside, some of the workers might be lounging outside where they could see the pens and watch the unloading going on. Then there would be Uncle Preston and one or two workers trying to capture the pig, while there would be a cheering gallery of spectators shouting encouragement to the pig or pigs as the case might be.

"Run, pig, run! You can make it! No, you goin' the wrong way!" And so on. Bets were sometimes made, and money changed hands.

Mostly the pigs failed to make good their escape. They were in too poor a physical shape to run long enough and hard enough to get away. A well-conditioned pig can run as fast as a dog for a fair distance and has tremendous traction, but these pigs had lived a life of miserable ease and could barely run a few yards before collapsing. Sometimes they did collapse and die.

Rarely would a pig complete his escape. Usually this was when a group would make a break and there were too many to catch.

Sadly, pigs that had successfully escaped to the woods and were then safe from men with knives would often come back looking for their dead comrades or perhaps for a handout. Then they would be recaptured. The workers called it "giving themselves up," as if they were escaped convicts.

There must be some deep lesson in all this for us as well as the pigs.

Many of the workers were on work release from prison. Like the pigs, they too would somehow manage to "mess up" and be returned to prison rather than freedom, often shortly before they were due to be put onto parole.

Some years before, a veterinarian had come to the plant to review the plant so it could be granted federal inspection. He was a black man, graduate of St. Paul's College, B.S., and Tuskegee, D.V.M., who had served in the military, and had taught school for a while.

Somehow, it fell Randal's lot to show him around the plant. He had been in high school at the time. Dr. Marion Adams had not been out of the service very long and was a bit timid about how to deal with the owners of the plant who were white and were almost certainly adversaries in matters of white versus black during the late sixties.

It was a hard assignment. He was to go and review a plant run by people who were not likely to be friendly and might resent his position of authority. He was normally a very likable, pleasant, level headed person. Not only might the white owners object to him, but the local black men who were very different in background and education might also resent him. Those who had sent him also may not have had his best interest at heart, but he ground through it all, polite, serious, and above all courteous.

Dr. Adams had grown up not very far away, in a small rural town on the Virginia-Carolina border. He knew the sort of people he would be dealing with. If they were going to make a success of the meat business, they would have to have enough sophistication to know that the rules he laid down came from the federal government, applied to everyone, and that they did not come from him, only through him.

When it was all over, Dr. Adams asked "Mr. Sharp, how did your uncles react to me? And more important, will they be willing and able to do what federal inspection requires?" In the back of his mind he was saying, "Do they mind having to listen to this here uppity nigger from Washington?"

Randal answered him with a perceptiveness that surprised him, for Randal was only seventeen and would go to VMI the next fall. "They are delighted to have you here. You are an example of what a black man can do and become if he has ambition, intelligence, and is willing to work hard in school." He did not need to add that money, community and family backing, were also needed and often not available.

Randal did not need to say that it was the poorly educated and poorly paid hardworking black men who might resent him. Somehow they might think that because Dr. Adams too was a Negro with ancestors who had suffered as theirs had he should do something to improve their working conditions. He would have agreed but could is not the same as should. He could only help them in the most indirect sort of ways.

Yet both Randal Sharp and Dr. Adams were very concerned about the conditions in the plant under which the black people worked. Even then Randal realized that coercion was not the best way to get good work out of people of any color or economic class. He knew that somehow the workers would have to believe that their success and the plant's were directly connected. In some ways it was not the most polite answer he could have been given, but it was honest. They would become friends in the years to come.

From time to time, Dr. Adams would return to review the plant. He would outline their gradual failure to catch up and keep up, although at this early time things were still looking up for the plant and the industry.

There was a very small sausage making plant in the area. They killed hogs too, although without the benefits of federal inspection. They had state inspection, which though no less rigorous in matters of sanitation, was more lax in matters of plant design and construction. Federal inspection was required to sell to the government or across state lines.

The sausage plant supplied several local restaurants with very fresh sausage. It had a very loyal following. Freshness was and is very important in pork sausage. It has to do with flavor rather than sanitation or safety. Of course, how long the product will remain good in the refrigerator or on grocer's shelf depends upon when the first day began, that is, the day the pig was slaughtered.

When the owner died and the little plant ceased production, Randal had then wanted to get their former customers by supplying those same small restaurants with very fresh sausage, too. "From hog to you in one day." Really it could have been overnight. Randal was thinking of a new process he had heard of using fresh carcasses from the kill floor, boning the meat hot and using dry ice in the mixers to bring the temperature down very rapidly.

Something else occurred to him much later. It would have eliminated the need for an extra cooler that would be built a couple of years later.

Sausage patties could be supplied to folks as far away as Raleigh, or even Norfolk, and it could be a really big business. Eventually it did become big business in the area and spread elsewhere. However, it was a decade later, and done by very different people. His father and his uncles simply could find little wrong with the idea except that no one else had done it, and that if they did it, they would need to have someone to ramrod the business and make it work. Finally if it did work, the big boys would be jealous and take it away from them. Besides, it was Randal's idea, not theirs.

A new kill floor was required to meet the requirements for Federal Inspection. It also would increase the production capacity of the plant. A few months following Dr. Adams' first visit to the plant, the new kill floor was complete. Immediately after the new kill floor was constructed with minor variations from the original plans to cut costs, federal inspection was granted. Much of the equipment was used and had been bought from auction sales from older plants in other parts of the country.

Even though it was less efficient than new machinery, it was still effective and met federal standards at a far lower initial cost. There were supposed tax advantages in repairing old equipment as a business cost over depreciating new equipment. Operating on borrowed money was out of the question to these men who so vividly remembered the great depression and the financial ruin of those who carelessly borrowed money during good times.

One of the things Federal Inspection required was the presence of a larger inspection staff, including a veterinarian present while the hogs were being slaughtered. Dr. Julio

Penario was a recent graduate of the school in Georgia like many veterinarians of that time. He had worked briefly in plants out west and had then been moved to Carolina and placed in charge of inspection in Sharp Brothers and Company. Dr. Adams was to be his immediate superior. Training schools for veterinarians and inspectors were just beginning then and training was still largely on the job.

There were also small plants nearby that might need him to visit on occasion about processing matters, particularly when there was no hog kill going on at Sharp Bros & Co.

Dr. Julio Penario believed the boss was boss. In the west he had come to understand that a plant had a normal chain of command that he could depend on. On the kill floor this meant the foreman was in charge and that if something needed to be done differently to meet inspection requirements he merely needed to tell the foreman.

The problem often was that the boss was not really in charge and that the man he talked to about specific problems had no real authority. An agreement made with a foreman would be overruled by someone else who had little understanding of the matter involved.

When the newly installed overhead chain was started, it was run for a short time, then turned off and then tightened with huge screw bolts. Chains wear rapidly at first and become longer in the process and so must be tightened. If they are not, disaster will strike when the chain is enough longer that the teeth in the driving sprocket fail to mesh into the slots in the chain and push out on the link. Then something breaks. Usually it is a weak link pin, intentionally made weak so it will break rather than some more expensive or dangerous part.

After a few hours of running and tightening the chain without problems, everything was pronounced ready for the kill the next day.

However, the next morning when hog carcasses were hung from the chain on gambrels, the chain soon broke. The entire plant seemed to shake as one might imagine a ship would if it ran aground. Having broken, the chain kept coming across the ceiling, freeing itself from the restraining blocks like a great arboreal python, until the motors were shut down. Had the chain been tightened regularly as had been done the day before, disaster would have been avoided.

Fortunately, the chain ran back only as far as the dehairer so that the other chain that carried hogs from the stunning pen through the scalding vat and into the dehairing machine continued to run. Dead scalded hogs were dumped on the floor. These had to be hung up and eviscerated by hand. This took several hours.

This strained relations all around. The plant managers had orders for meat while the workers and inspectors had not planned on a day which might stretch well beyond eight or even twelve hours.

Dr. Penario thought this was a very poor beginning for his first day at Sharp Brothers and for their first day as a Federally Inspected packing plant.

Meanwhile, the chain was repaired. By running only a few hogs at a time and reducing the weight on the chain they managed to kill the planned number of hogs. It became a very long miserable day.

Randal did not get to go on the date he had planned for that evening and was not even able to call the young lady to call off the date and tell her what had happened. It broke up their friendship. Not that she did not forgive him. It was that he could make no plans, his life was not his own, and he could never say "I'll be there at six, or eight," and know he could keep his word. The plant seemed to smooth over its troubles. Randal went to VMI. He had been hazed by experts before he ever got to Lexington. Inwardly he had become very tough by the time he graduated.

There were other exciting events which should not have been. A storm cut off the electricity. Randal had long before asked what they would do about a power failure. The reply was that they would figure it out when it happened. Getting an auxiliary generator was out of the question as too expensive.

When it came, there was a mad scramble to pull hogs out of the scalding vat by hand. Hogs that were left in the scalding vat for a longer time than they were supposed to remain would eventually parboil and with the viscera inside would have a sort of foul odor. They would have to be condemned. Randal was away in school at the time but heard about it and quietly cursed the stupidity of it all.

So the hog carcasses then had to be lowered, from the vat to the floor, scrapped and eviscerated by hand. A mad scramble for ice came to naught. Fortunately, after an hour or so the electricity came back on. Soon it was arranged to have two power lines from opposite directions and different power plants coming into the plant. But why, he wondered, wasn't it done before there had been a costly disaster, rather than after?

The final incident for Randal came after Dr. Penario and the processing inspector were looking at hog carcasses in the cooler. There were several carcasses with fecal contamination on the feet or the back, or worse in the meat in the neck where the head had been cut off. It was something that had gone on for a long time. The idea was that it could be trimmed off at the rail station by the company trimmer. Uncle Bo simply insisted that the rail inspector and the company trimmer should do a better job. Dr. Adams had objected that carcasses were supposed to be ready to go in the cooler free of contamination or defects before they even got to the rail station.

It happened because the eviscerator's knife sometimes cut into the intestine, spilling intestinal contents onto the hog's feet as he dragged out the viscera with one hand and cutting them loose with the knife in other hand. It was an indirect effect of the Packers and Stockyards Act. Animals going to slaughter were often fed and watered heavily before going across the scales. Feed and water cost much less than meat but brought the farmer the same price on the weigh master's scale.

Finally, Dr. Adams told Dr. Penario to rigidly require that carcasses were to be cleaned immediately after evisceration. That meant stopping the chain and trimming with a knife, not just washing off dirty feet. Stopping the chain slowed operations along the entire production line.

"After all, we aren't poultry inspectors!" Dr. Adams declared. He had been involved in the shotgun marriage of Poultry Inspection and Meat Inspection. Poultry inspectors allowed fecal contamination to be washed off and spent a lot of time being sure that a plant that killed chickens or turkeys used enough water. In red meat plants such contamination had to be trimmed off with a knife. Red meat inspectors suddenly placed in poultry plants were horrified at the practice. Of course, the best answer was to make such contamination so rare that it would have not have mattered had the entire carcass been condemned.

Uncle Bo had a fit when Dr. Penario explained what would have to be done. Eventually he was made to understand that it was better to trim one carcass rather than three, for the carcasses on either side of the one with dirty feet invariably got contaminated too. But that took months to sink in.

At the moment Dr. Penario stood along side of the viscera inspector and stopped the chain whenever he saw a dirty foot on a hog carcass. "You can't do this to me." shouted Uncle Bo waving his cane.

Fortunately, Dr. Penario was out of reach.

"What do you mean I can't? I just did!" was the cheerful and perhaps impudent rejoinder. Meanwhile, thirty employees were being paid for standing still while one or two trimmed the contaminated carcass.

Dr. Penario explained the what and why of what had to be done, but Uncle Bo was too upset to understand or listen. If anything, it should have been done long before and Dr. Penario was wrong in ever having allowed it to have been done differently.

Uncle Bo was shouting and waving his cane. "You can't come in here and tell us what to do! We built this plant. Brick by brick. We know this place and our business. This is OUR plant!" He went on for what seemed several minutes.

"Perhaps not, but we, I, can withdraw inspection if you don't do it the way we in USDA require. And you will have a hard time selling meat without federal inspection," stated Penario, raising his voice only enough to make sure he was heard.

"Probe as long as you hit mush but retreat when you feel steel," said an old proverb. Uncle Bo had hit steel and he retreated, angry and rebuffed. The young veterinarian was called to account by Dr. Adams as to whether somehow the required immediate cleanup of hog carcasses could have been handled more diplomatically.

Julio bluntly said, "No!" It probably was true.

It had happened shortly before the Christmas holidays. Randal and his mother had sat up watching Charles Dickens' Christmas Carol the night before. As Randal watched Uncle Bo waving his cane at Dr. Penario and shouting about how it was his plant, how he had built it, painfully, brick by brick, Randal realized that it was almost a parody of Marley's ghost describing the chain of cash boxes he had been condemned to drag through eternity. Randal broke out laughing. Later when he had time to think, he realized sadly that it was true.

Uncle Bo had never married. It was not that he lacked charm, for he was a very different man when he was away from the plant. Suddenly, Randal realized that his uncles and his father actually hated the plant. It was a tyrannical master. If it had given them riches, it had taken most of their waking moments. It ran them rather than the other way around. Because of the great depression, they had been robbed of the college education he and his cousin had gotten. It had set them against each other so that each feared that somehow the other would cheat him. They and the plant were at war with each other. It was a great revelation. He said only a few words to his cousin who was a few years older.

"So you know," remarked his cousin, Horatio's son. They had a very frank talk. "I'm trapped like Dad and Uncle Bo. I'm nearly forty with a wife and child. I've got to stick and hope that I can turn it around. I can't do that while the old men still set policy. I just do what's necessary, draw my pay, try to save it, and hope for the best. I gave up trying to change things years ago."

"You won't have time. It will have gone down so far by the time you are the boss it will be too late."

Randal Sharp left and did not see the inside of a packing plant for the next thirty odd years. He had already graduated from VMI, and finished his MBA that following spring. When he left the plant for good, he served briefly in the military and afterwards went to law school at UVA. Then he went back into law enforcement with the military.

Randal suggested to his father, Preston, that he would be smart to sell out his part of the plant and retire. Perhaps he took his son's advice because a year or two later he did sell out his part of the plant. Another decade saw the plant sold to one of the larger concerns in the area at a fraction of the price it might have brought had the owners kept up to date in their business and personnel practices. A decade after that, the plant was torn down, because of obsolescence.

Randal Sharp enters our story again, much later, but this time as a senior police official.

Beyond the Pangs of Remebered Grief

The night seemed crisp and very cold when Dr. Heinrich Kruger opened the kitchen door, crossed the porch, and went out through the snow to the garage. There was only starlight for it was long before dawn; only starlight glistening on the snow. It was the winter of 1976.

He opened the garage doors. Inside, the Volkswagen was clearly visible. The thermometer on the wall said minus twenty. He wasn't quite sure what that was in centigrade degrees but he knew it was very cold.

Later he would think about it. Surely Stalingrad had been colder.

"Ah, yes, the name had been changed," but since he generally thought in German, it was;

"Ach, ya, der Name wexelt war. Nun es ist Volgagrad."

"Ah, yes, the name was changed. Now it is Volgagrad."

It seemed that even if Stalin deserved demotion, a place of such heroic battles, of such tragedy, should have kept its name. But then perhaps it is better to forget. Forgive and forget.

He opened the car door, seated himself, and turned the key. The engine cranked reluctantly, chuffed a few times and then began to run in earnest.

"Naturally, a German car is reliable," he muttered.

He got out, unplugged the electric wire from the engine heater, wrapped it up, got back into the car, backed out, turned around in the powdery snow and headed for the village where his packing plant waited for him. He had to be there before operations began and it was thirty seven miles, thirty seven miles through the cold starlit Wisconsin night. Beautiful but deadly, for if the car went off the road, or a sudden blizzard came, he would be forced to remain in the car without heat. For this he had made adequate preparation, food, blankets and a catalytic heater. Colder than his native Germany, perhaps as cold as the Ukraine, he thought.

Behind him were great cloud banks, looking like great threatening mountains. These would bring more snow during the morning, perhaps by daylight. Long before then he would be at work in the plant.

In the brief time allotted to the trip, he would let his mind roam through his past. Here he was, a German veterinarian, in a German car. He could almost imagine that things had been different. He could go almost all day, at home, in the village, or even at work speaking only German. It was a different sort of German dialect, but acceptable. If only he had his family with him. That was a hard thought which he put out of his mind. There was now only his older sister who lived with him. Her children and grandchildren had grown up and gone away. She was a widow. There had been an older brother as well, but he had died at Verdun.

The first time he had ever seen a Volkswagen was in a Hamburg showroom. He had made a first payment. Then the war had come and taken him away from his civil service position as a veterinary meat inspector and placed him on the Eastern front as an Army veterinarian.

He had once applied to join the NSDAP (National Socialist German Workers' Party) but in the shuffle his papers had been misplaced and he failed to be accepted. What he had first considered bad luck turned out to be good luck for it had allowed his entry into America.

For a while there had been another kind of Volkswagen, a military model with a sergeant to drive it. Most of all there had been endless numbers of horses. Most needed tetanus toxoid and shoes. Later there were horses that needed their wounds treated. What he would have given to have the sulfas and penicillin veterinarians now so liberally used! Finally as things went bad, there were the horrid decisions; treat the animal if it could be saved, or kill it for desperately needed food, or declare the animal in such bad condition that it was not fit to eat.

When he arrived at the plant, he had two visitors, a farmer and his young son, a boy of perhaps thirteen.

"What can we do for you?" asked Dr. Kruger. It was not unusual for farmers to want to follow their hogs through the slaughter line to observe for themselves how carcasses graded out, or the results of a worming program. But this was different.

"Doctor, we loaded a young boar on the truck by accident. He wasn't meant to come here. He was our new replacement boar, a grand champion, and quite valuable."

"Please give us back Sparky," pleaded the boy.

"That's enough, son," said the father, a hard faced man in his early forties. It was obvious that such a show of emotion seemed unmanly to him.

"Let me see. You have unfortunately loaded a young boar onto the truck with the other hogs being sent here for slaughter and you believe he is here. Can you be sure of that? Let me explain. The pig doesn't belong to me, or the government. He belongs to the Weiss Packing Co. And to get him back they will have to agree. Then I will have to get permission from my superiors in Madison. They in turn will have to ask the APHIS field service people for permission for the boar to go back to the farm. They will probably allow it since there have been no recent outbreaks of contagious swine disease recently. If there were an outbreak of perhaps, ah, (schweinpest, he thought quickly) hog cholera, it would be out of the question."

The father thought for a moment and said, "Well, we loaded the pigs into two of Johnny Doyle's possum belly trailers yesterday and Sparky turned up missing last night. We had some small sows in there too so we could have missed seeing him. He wasn't much bigger than a big market hog.

"We paid a lot for him when he was hardly more than a feeder pig, and he is sort of a pet. Joe, here, sort of raised him this fall and we figured he would be big enough to use by spring. He was worth driving all night to try to save."

It was then that Dr. Kruger could see how tired they both were. In the harsh light of the office he could see that the boy had been crying, but now the boy was choking back the tears.

At least the office was warm.

"We must hurry then. Can you identify him? What kind of pig was he? What does he look like?" asked Dr. Kruger.

"Why, he is white, sort of long, like any good Yorkshire boar, with ears standing up and an alert face. Friendly too. Would always come when we called him," said the boy.

SPARKY

There was something familiar about the boy's appearance or manner but Dr. Kruger couldn't be quite sure what it was.

The telephone rang. The veterinarian answered.

"Kruger here. Good, yes. We will be right there." He turned to the man and his son and continued, "So you have already talked to the Weiss brothers and have explained your problem. The plant people are willing to give, I should say sell back the boar if you can find him, and if the authorities agree to let him go back. But first we must find him. It will be some time before we can contact anyone in the Madison office. In any case, we must

205

begin antemortem inspection now. We can't cause the plant any delay in their operations."

The three of them walked out to the hog pens. There, one of the men who worked for the plant would drive the hogs out of the each pen, parade the hogs forward and back so that they could be "seen on both sides, in motion and at rest."

The weigh master, Mr. Johnson, met them. Dr. Kruger spoke to him first. "Mr. Johnson, this is Mr., sorry, I didn't catch your name, sir."

"Sorensen, Ben Sorensen, and this is my son Joe."

"They told me about losing your boar. Too bad. He may be plenty hard to find. But we'll try. You say they came in last night, on Doyle's trucks. They would be in pens 7, 8 or 9, I think. But we may have killed some of them yesterday."

The last remark seemed to panic the boy, but his father put his arm around the boy and said, "We aren't beat yet. Let's look before we get shook up."

Mr. Johnson continued, "We will run the regular market hogs first. You can look at them too if you want. Just be careful and don't get hurt. Well, you know how hogs are. Just don't get run down."

The hogs from the first pen were run out, forward and back, and then into the alleyway leading to the shackling pen. They all three looked carefully but none were anything like Sparky. They were obviously too small, and as well as they could tell none were boars. Boars were not supposed to be killed on the regular hog kill, although sometimes they were when the men running the hogs were less than alert, or lucky, for young market size boars were sometimes hard to spot. But if they were missed, it was very seldom that there was any hint of boar odor or taste.

They were simply too young. Boars that go through were supposed to be retained for closer examination, the testicles being placed in the pan with the rest of the viscera. Dr. Kruger had what he considered a certain method of identifying boar taint. He put his mouth where his signature was. He and the kill floor foreman would cook and eat a piece of tenderloin from the boar. It they found it without taint, it passed. If tainted, it was condemned.

A second and a third pen full of hogs was sent. As they watched the third group. Dr. Kruger called out, "Hold that one!"

There was indeed a large bulge in the rear of the hog he had pointed out. But it was due to a scrotal hernia. It had been castrated, but the intestines had managed to get into the scrotum. Dr. Kruger looked at him carefully, and then said, "Let him go on, I don't think that will cause us a problem."

When they ran out the fourth pen, again Dr. Kruger called, "Get me that one. Yes, that one with the lump."

But the hog didn't want to be gotten. It didn't want to be separated from its companions. Mr. Johnson had to use his cane and his shock stick to keep the pig from following the rest of the pigs down the alley to the shackling pen.

"Put that one in the suspect pen."

The others could see what was wrong with that one. It had a large lump on the side of one ham.

"Probably an abscess," said Dr. Kruger.

"Sure looks like it," admitted Mr. Johnson.

While Mr. Johnson was putting the hog into the suspect pen, Joe asked Dr. Kruger why the hog didn't want to be left behind, especially since certain death lay at the end of the alleyway.

"I used to think it was just because they don't know what is happening over there, which I'm sure is the case, but I think there is something else. Sometimes when we need to separate the hogs from each other they make a disturbance, and fight us.

"Sometimes I think they are loyal to each other, just as we are. After all, they grow up together, even though we shuffle and sort them here and at the markets." He was about to say more but thought better of it. "I have seen men go to certain death to be with their comrades, when escape was possible by deserting those comrades," he thought.

By now, the kill floor was in full operation, with hogs, or rather their carcasses, reaching all the way into the coolers.

If a carcass were retained, a call would go out for Dr. Kruger to come inside and look at the retained carcass and viscera. But so far there was none and they continued to look at the hogs. Still no sign of Sparky. They looked at pens full of hogs, called out, "Sparky, Sparky," but no hog came.

Now they were looking at hogs that were being put back into pens. Dr. Kruger would look at his watch, sign the pen card, write down the time, and hand the card to Mr. Johnson. Soon all of the smaller market size hogs would have been seen. Still no Sparky.

A call came for Dr. Kruger to examine a retained carcass. He went into the office, changed some of his clothes, made a telephone call, since it was now late enough to reach someone at the main office in Madison. Then he wrote up some papers; kill floor forms he called them.

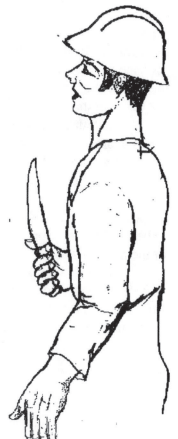

208

He talked briefly with the processing inspector about what was clean and what wasn't.

Then he went inside and examined the hog. It had large red diamond shaped splotches on its back. He cut down into one of the blotches. The red color extended all the way through the back fat into the muscle below.

He looked at the heart on the viscera pan, cutting it open but found nothing amiss there. He looked at the lungs. Part of the lung was white and spongy, part of the lung was soggy and dark red. He looked at the kidneys which were still in the carcass. He removed one and, having first peeled off the outer membrane, held it up to the light and examined it carefully. He liked to be thorough even though he had decided to condemn the carcass almost as soon as he had seen it. With his knife he made a series of X shaped slashes on the carcass. Then he dipped his knife once more into the sterilizer, a basin of almost boiling water, and replaced it in the metal scabbard he wore. On a clipboard beside the sterilizer he wrote, "Swine Erysipelas, bright red diamond shaped skin lesions, pneumonic lesions in lungs, petechial hemorrhages on kidneys," on a brown government form.

He placed an X and the code number 912 beside it. He reached over into a small pan containing a metal stamp and an ink pad, took out the stamp and stamped the legend "US Inspected & Condemned" on the carcass.

Hardly had he done this than another carcass was railed out to him. It too, he stamped with the same stamp, without even looking at the viscera. The carcass was a bright yellow color.

Then he took out his knife again, and almost as a afterthought thoroughly examined the viscera. The liver was larger than normal, far harder than normal, and instead of a dark brown mahogany color, it was literally a bilious yellow. At that moment Ed Weiss, one of the brothers who owned

209

the plant, and who served as kill floor foreman arrived along with his helper, Harry Jones.

"Er trank zu viel Schnaps, nein?" He remarked, laughing, as if the pig had any chance to drink anything other than water.

"Let that be a lesson to you, Harry."

Which was something of a joke because while Ed and his brothers drank a bit, Harry didn't.

While he didn't like losing a hog that had cost nearly fifty dollars, he was in total agreement as to the disposition of the carcass. He wouldn't have passed either of the two pigs. In fact, it was seldom he differed with the veterinarian as to what he considered fit to eat. If anything, his standards were harder. Some things looked bad but might not be condemnable and while the good doctor had more technical knowledge, he didn't put himself in the place of the customer like Ed did. Besides Dr. Kruger would sometimes say; "Ah, but much worse we had to eat during the war."

Ed Weiss would reply, "Leider es gibt kein Freibank heer." In Germany, meat that was not up to the best standard, but still safe, was placed into a special category referred to as the Freibank, the free bench, and sold for a lesser price. America however has only a single standard of wholesomeness. Harry cut the ham strings of both hogs, dropping them both into a large two wheeled cart with "US CONDEMNED" painted in red letters on both sides. Then he poured a green fruity smelling dye onto the carcasses. Then he carted them out to the tank room.

Meanwhile Dr. Kruger wrote up the second hog just under the line for the first hog.

"ADX223133 Icterus... Bright yellow color, cirrhosis, X 540"

As he stepped down he felt a slight twinge of pain in his stomach, a sort of cramp. He stopped a moment and it passed.

Now it was time to return to the pens and check out any more hogs that had come in. Also, there would be a good bit more to do if Mr. Sorensen and his son were able to find Sparky. It somehow seemed unlikely that they would. He would have to make those phone calls to Madison, fill out papers, etc. It would mean extra work, but he hoped they would find Sparky. Something about their quest touched his heart.

Back at the pens, he found Mr. Sorensen and his son. They had not found Sparky. Nor did it seem likely that they would. They had looked in all the pens and at all the hogs that might have been Sparky. All the hogs had come in that were coming, or at least all of the truckloads that could have contained their pet boar had arrived and been unloaded. Neither could they be sure that they had not seen Sparky and just not been able to recognize him. But they had called out his name and there had been no response. Had he ignored them? Did he fail to hear their call? Who could say?

Finally they thanked Dr. Kruger for his help as well as thanking the weigh master, and despondently gotten into their pickup truck and driven away. Somehow it still seemed dark and snow was falling. The storm had caught up to them long since.

There was a brief persistent gentle tugging on the hem of his coat, as if a small child were attempting to gain his attention. When he looked down there was a pig nibbling at his coat as if to say, "Please sir, could I have a few words with you?"

But as soon as the pig had gotten his attention it was embarrassed and turned and ran. As it did so, he could clearly see that it was a female. It was not Sparky. Dr. Kruger stood alone in the pens and talked to the pigs. He said to himself softly, "Are you here, Sparky?"

There was no answer. He thought he remembered the boy's face. How like his own son he seemed now that he thought about it. His son had died at Stalingrad or at least he had never come back from there. He himself had been there but had flown out before things had gotten really bad. He had heard that his son had been wounded. When he was able, he went from one field hospital to another to look for him. He had the chance through another army friend to take his son with him in the evacuation, if he could find him.

Sadly, he could not. If his son were at any of the places he visited, he did not reveal himself. So flying away from Stalingrad he left the last German member of his family. His wife and his parents had been killed in the fire raids on Hamburg. He was the sole survivor. His sister was an American, and there had been a very long period of separation and alienation caused by war and politics. Her sons had fought Germany and Hitler, and were not sure they wanted him for a friend.

Thinking about it all, especially in the deep dark days of winter filled him with grief. It was a grief that would come and go but which never really left him. He wondered if life could have been any different. Was there anything more to life than just work, and waiting for the end of life?

Surely there must be. He had discussed his life with Pastor Justis. He had a good friend in the pastor. He thought he had been as good a man as perhaps circumstances permitted. Certainly he had not taken part in the horrifying atrocities of the SS in the Ukraine.

Still it hurt to think of the wastefulness, the useless sacrifice of so many, especially his family.

He went to the small office adjacent to the weighing scales. There was no one there. The weigh master had gone up to the main office for a time, perhaps to an early lunch, or to settle accounts.

He still could suffer the grief of the small boy for his porcine friend. It was like a small reflection of his own. He felt a lump in his throat. A small pain that began in his throat, that seemed to radiate downward into his chest, a pain deeper than the pangs of remembered grief, that filled his heart. He sat down at the desk and fumbled for the nitroglycerine he carried. He put the pill under his tongue, put his head down, resting it on his forearms and waited for it to take effect. "Lieber Herr Gott, bitte vergeben mir alles," he thought. "Dear God, forgive me."

Suddenly he felt much better, better than he had in many years. He stood up, he looked down. Someone was sitting in the chair resting his head on his arms, an old man in a white coat, with rubber boots and a white plastic helmet beside him on the desk.

There was a knock on the door. He looked at himself. He was in uniform; in the fieldgray dress officer's uniform he wore when going on leave. The door opened. It was Sergeant Vogel. Outside was a military Volkswagen. Sitting in it was his elder brother. None of this seemed out of place but there was something or rather someone missing.

"I had thought Franz would have been here too."

"No, Herr Kapitan, he is a colonel in the DDR."

So his son still lived!

213

When the weigh master returned, and could not wake him, they called the rescue squad, but they were much too late. They buried him in the Lutheran cemetery beside the church.

Years later they still spoke of him, saying that he was a good and kind man. As for the boy and his pig, it turned out that Sparky had tunneled under a fence and had gone on an amorous visit to a neighbor's farm. There he had been recognized. Mrs. Sorensen had been called but had not been able to call the menfolk to tell the good news.

Bloody Instructions

Winslow came on duty a little before midnight. There would be the usual midnight to morning shift consisting of a seven hour hog kill. He would be the lone Government Inspector present. There were two specially trained plant employees who would do much of the actual examination of the hogs.

The plant was located in northern rural Alabama.

In the government office there were just three chairs, a file cabinet, and a desk with a computer sitting on it. Winslow turned it on, and logged in. The date appeared; April 27, 2014, the time, 23:47. In a few minutes it would be midnight and the beginning of a new day, April 28.

First Winslow checked the log, how many animals were to be killed, where they came from, and other pertinent data. He had already given them a once over with the thermoscope. Any hot spots on any of the animals would show up as a red patch on an otherwise green hog.

The hogs were 300 lb. Kalbert Klone 467. All supposedly as identical as any identical twins. By breaking up the morula of the embryonic pig into many cells, and repeating the process several times, thousands of identical pigs could be grown. Since the same company reared the hogs as slaughtered them, there was no reason to fill them with feed to fool the weigh master. Nevertheless the hogs were weighed in individually to determine how much deviation there was in individual weights. Any hog that was more than 8% lighter or heavier was of interest to the company and the Inspection service. Automatic computerized scales made the task easy.

An electronic transponder in each hog gave back a signal when queried by radio with the hog's serial number and other information. This remained with the hog until it was a carcass to the cooler, where the transponder was removed, eventually to be placed in another hog. This implantation of the transponder was the only surgical procedure ever done to the pig. Castration, ear notching, tail removal, as well as nose rings had vanished as a result of biotechnology. Clone pigs were all of the same sex, usually male because they grew a little faster. They were not castrated because boar odor had been bred out of them, and would not develop until the pig was much older, perhaps at ten to eleven months. Tails had been bred out as well, as had the disposition to bite each other.

Unlike pet mini pigs, these giant hogs were not very bright and were lacking in personality. It was also said that the life span was shorter because of the cloning process and the Hayflick limit. There is a limit to how many times normal mammalian cells can replicate, so that the number of times cells were replicated to produce identical progeny reduced the number of times the cells could replicate to counter normal aging. However since they reached 300 lbs at four and a half months no one cared that they would live only five or six years if kept rather than slaughtered.

It had not been practical to breed hairless hogs. It had been tried and it was found that the bristles provided necessary skin protection.

So hogs were still dehaired by putting them in a scalding vat and then beating the hair off of them. This was still done with a mechanical device using rubber composition paddles which beat on the carcass hard enough to remove the loosened hair but not hard enough, the packer hoped, to break the skin. It had been found that the hair growth could be regulated by the amount of bright light the hog was subjected to so that the hog would have a late winter coat that the beaters could easily pull out. This avoided the skinning that used to be such a large part of dressing a hog at certain times of the year. It also made the skin usable as food.

Perhaps it would be a nice quiet evening. There would be a little paperwork to do. Paperwork was not quite the right word for it since his report would go directly into the computer and thence via modem to the regional office. A printed copy was also sent.

Normally, there was a good bit of reading to do. He could either read it directly off of the computer screen or print it out and then read it. The government preferred that he use the screen and save the paper.

"Oh," he thought, "tonight is the night for the rebroadcast of Il Padrino. I bet I've already missed the introduction." He was well advised to try to hear it soon for it might be placed on the list of prohibited works because of its references to violence. Violence in literature was generally prohibited as it was on screen and stage. Perhaps he could get a bootleg copy to keep. He already had a copy of Tosca which he believed might never again be performed.

He kept it quite secret. Perhaps if they knew they didn't really care. Still, some enemy might denounce him and that would mean trouble.

He remembered when the gun police had broken into his grandfather's home to take away the family shotgun that had hung over the family mantelpiece for so many years. Family legend said it had been used to prevent night riders from taking his great uncle away and lynching him. But that was a long time ago. It probably wouldn't matter now. There were video cameras nearly everywhere which recorded where people were and what they were doing.

He adjusted his headset that was attached to his helmet. People wondered why he wore his hat in the office, but telephone calls came through his headset, along with other messages from within the plant. When someone wanted to talk to him, the message overrode the radio breaking in and interrupting his listening. He liked opera. Someone else might be listening to a ball game. It made the physical work go faster. Unfortunately, this would be a night with major interruptions.

The voice in his helmet spoke, "Following the Corelone overture the curtain rises. Outside, off stage one can hear sounds of a party, a wedding celebration for Don Vito's daughter who is being married to the Don's godson, Don Carlo. Don Vito is alone sitting on his desk. The off stage clamor fades away and Don Vito sings a soliloquy, in a low Sicilian dialect.

"The words are a paraphrase of words and thoughts taken from Niccolo Machiavelli. The tune is from the film of some forty years ago by Nino Roto. In fact, the opera trilogy owes more to the films than to the original book by Mario Puso.

"This scene is repeated at the end of the final opera but this time it is Don Micheal who is singing a sardonic version in classical Italian, and ad-libbing in English, that the world changes and yet remains the same. It is true for events have come full circle and Michael Corelone is now Il Padrino."

Suddenly the announcer's voice in the head phones was interrupted by another voice. "Mister Winslow, will you please pick up the handset and turn on the secure line."

"Now what?" he thought. Then back to the opera.

"Soon Amerigo Bonsera enters. The others fade out of the scene as he and the Don sing a duet. Bonsera describes the insult and injury to his daughter and begs the Don's help in seeking revenge. Don Vito asks why he should help Bonsera now since he has had no dealing with the him previously. In the end, however, he agrees to avenge the daughter's insult and Bonsera is calling Don Vito, Padrino mio."

"Hello, is that you, Winslow? This is Gonzales, Alfred Gonzales, from compliance and security in Nashville."

"Yes, I am Erazmus Winslow, the inspector here for this night shift. What can I do for you?"

"Can you give me your license number?"

There was a long silence. The license number was used as a password, to prove identity, sometimes along with a social security number. This was something serious.

"I take it this is serious business."

"Indeed it is."

"Something illegal?"

"I'm afraid so. No, not something you did. At least we don't think so."

Winslow gave him his number.

"You must hold up the kill, use a pretext if you can. It will take a while for someone to reach you, but they have been on their way for a while and should be there before long. I guess I can tell you over the phone that when you pulled the kill floor program last Friday, we checked it and found a discrepancy. We have to find out what is going on."

The machines on the kill floor were all computer controlled to some degree, even though there were a dozen employees still utilized. If something had been changed on the program directing the machines it had to be checked and justified to the inspection service.

Winslow sat thinking for a moment and then said, "I'm afraid I'll have to give the news to them straight. Is there some reason I can't? After all, computer programs sometimes develop bugs that could cause real mechanical troubles without there being any skullduggery afoot."

"Can't you find a little dirt out there somewhere?"

"I suppose I could find something if I think about it. There is a light out that I can insist on having replaced."

"Okay. Just do it and perhaps our people will arrive shortly."

Winslow hung up and put his helmet with headset back on. Then he took a tag and hung it at the shackling pen.

The opera prologue resumed. " 'Tom Hagan is my name and I'm a private lawyer for a very important client.' He sings in a lilting Irish reel. He continues in this pleasant bantering way for a few lines. But soon there is a subtle shift in the rhythm, the key changes from major to minor and we recognize the sinister tarantella of Don Vito as Tom Hagen begins to present the Don's requests."

Hardly had Winslow hung the tag when out came the foreman who took one look and headed back towards the repair shop. The foreman returned with the electrician in tow carrying a large bulb. They knew enough not to argue when the failed light had been on report for several days.

It took only about five minutes to replace the light. Meanwhile, Winslow was looking for dirt. There were usual places but it turned out that they were clean. The automatic cleaning devices when properly serviced and maintained did a very thorough job. Surfaces were covered with a fluorocarbon plastic that shed water and dirt, and were arranged so as to minimize dirt accumulation. Packers knew that meat that was washed down drains could not be sold, so saws became rotary knives that shed very little meat "sawdust".

Even knives had improved over the years, being made of two laminated layers of stainless steel with a layer of tungsten carbide in the middle. Like a beaver's tooth it was self sharpening, although it had a brittle edge that required careful treatment.

The foreman, whose name was Mike Long, and the electrician finished with the ladder and the light and both turned to Winslow and in one voice asked, "Now may we start?" It was strange to hear their voices both inside the headset and around the earphones as well, or so it seemed. It was really an illusion, since the ambient noise would have drowned out normal conversation, but he could clearly see their lips moving and hear their voices.

Now he feared he would have to give it to them straight; no kill until the computer program was cleared up. Computer programs were sometimes altered to run the line at illegally high speeds, or to skimp on the amount of water used, or the amount of light throughout the plant, or to cut carcasses in ways that increased the size of an expensive cut at the expense of a cheaper one. The maddening thing about that was that the inspection service really didn't care so long as the meat was wholesome, but the meat industry did. Labeling definitions had made the inspection service act as policeman so that exact definitions of each cut were required and adhered to.

Winslow gave a long sigh, looked at the red bandanna around Mike Long's neck and wished he were back in the office, listening to the opera and reading.

He was about to begin, when the visitors were announced.

"This is Staff Director Yardley. Mr. Winslow, are you holding up the kill?"

"Yes sir, I am, but I was just about to release the kill floor to begin."

"Well, hold the kill until they the reinstall the previous program, or they explain this program you sent us last week."

About this time the man Winslow was talking to came onto the kill floor. With him was another man. Director Yardley introduced himself. "Permit me to introduce myself. I'm Dr. Yardley and this is Police Colonel Randal Sharp," as he introduced the other man.

Both the electrician and the foreman were rather shaken by the appearance of the two officials. So was Winslow. Especially since one of the men was a high ranking policeman in the National Federal Police that had replaced the old less efficient state and local police and operated as a fourth branch of the military with previously separate federal agencies like the FBI and CIA included in it.

"I'm the kill foreman here. My name is Mike Long. Does this problem involve the kill floor program?"

"Yes it does."

"Well so far as I know, the last update to the kill floor program was issued over six months ago and seems to be working well."

"That seems to be the problem. That last update doesn't match the program Mr. Winslow pulled last Friday."

Pulling a program involved making an extra copy which was filed in the government office and kept there and checked against the program in use to see if there were any differences. Sometimes they happened due to wear and tear on the disk. Really big programs used a laser disk. Usually there was an additional magnetic area on the disk for small changes. It was here that changes were made. Often subroutines for individual machines were imbedded in the laser disk, with customized coordinating routines between machines on the magnetic side of the disk. It was these that had to be sent in via secure modem to be analyzed and permitted or rejected. Such small additions were only allowed for a short time and then had to be made part of the master disk which was optical.

Winslow had done a comparison with the previous program, found a difference, and submitted the program, expecting to receive a routine clearance for it. He should have been notified by the plant programmer of any pending changes but he had not been. A minor infraction, he had thought.

Then the foreman asked whether they could use the previous program and get on with the operations. Neither man had any objections. They would have to be sure that it was the identical program. That meant running a duplicate in the office of the older program in the file in the Nashville office which would take several minutes, even with fiber optics and advanced computers. Then they could also compare it with the program Winslow had in his government file. There was a small chance something might differ there as well. Winslow did not ask how his file program might differ from the official Nashville file. It would have been unlikely that it could have been changed without his connivance.

Silence lasted only a moment. The opera description returned to Winslow's helmet radio.

"There is a knock at the door. Six men enter carrying a what appears to be a huge gray rug on their shoulders. When they lower their burden onto a large table, it turns out to be a huge newspaper with the headline, 'Gangster missing.'

"Inside the newspaper is a shark. The men are singing a dirge based on 'Many Brave Souls are Asleep in the Deep'. Luca Brazzi has been killed. They sing of brave and terrible deeds using horrifying gestures indicating how Brazzi killed the enemies of the Don."

For a brief time, Inspector Winslow and Director Yardley went back to the government office leaving Colonel Sharp and Mike Long alone on the kill floor. The police officer remarked that he had worked in a packing plant many years before but that things had changed in the intervening thirty odd years. Despite the very long interval since he had been familiar with packing plants, it had something to do with his choice for this particular investigation. He asked the foreman to explain the working of the plant to him.

Long began, "This plant is some fifteen years old and some of the techniques here were experimental. For instance, we use a turntable system here which has generally been abandoned for parallel lines which were what existed before. You see that this is a pretty compact kill floor. It was intended that men with machines to help would work this kill floor. So you see the four rotary work platforms with the chain winding around them.

"The stairs to each platform are located where the chain runs from one sprocket to another. Those sprockets are from sixteen feet to twenty feet across. The chain runs at eighteen hundred hogs an hour. The platforms turn at about half that speed, so that the workers can keep up. No matter how many people you have, they can't read the writing on a freight car if it's going too fast, but if you can run alongside, that is a different matter.

"The sinks and sterilizers are located in the center and don't turn. We often get asked how we keep from winding up the pipes and hoses. Actually they don't move. The platform rotates around them."

"Since we have some time, can we see the rest of the plant starting with the pens?" asked Colonel Sharp.

"I suppose so. I see you already have a coat and helmet with a communicator," responded the foreman. They walked out a side door, down a lane, and soon were looking at the unloading pens.

There seemed to be fewer pens than Randal had expected. There were low sinusoidal walls about two and a half feet high. Overhead were rails and lights. Somehow it looked like there were no nooks or crannies for junk to accumulate in.

Neither did there seem to be much space for trucks to park and be lined up to unload hogs across the scales, each truck taking its turn. He asked about having enough space for hogs for one day and was told that no one had provided enough space for a day's hog kill since he was a boy. The regulations and practices had changed. In an emergency, hogs would be off-loaded in one of the pastures or farm pens.

Sharp had been unaware that the hogs were raised very close to the plant and could generally walk to the plant without being carried on trucks. After all it was dark and the other buildings were not easily visible. He had probably read about it but forgotten. It was very different from the packing plant his father and uncles had owned forty years before.

"We could go back a mile or so and see the whole thing. Of course we couldn't let you into the sterile areas where we raise the baby pigs. But you could see what videos we have and could see via remote if it were necessary."

Seeing the pig raising facility could have been necessary under other circumstances. USDA and FDA had done such watching for illegal drugs and for what were called "cruel practices" under the Animal Rights Act of 2008. Generally it involved special police rather than simply the criminal police force to which Colonel Sharp belonged.

The hogs were exercised and moved regularly. As they grew they were moved to larger facilities to accommodate their increased size. It involved a sort of training to move when required with rewards of special food at the end of the trip. One of the tricks was to play a special tune each time so the pigs would get used to being moved. Usually it would be a march. Eventually it would end for them in the killing pen, but they would approach the end without fear, not expecting the final end of the road. It was done this way because it worked best economically, rather than any soft hearted reasons as to what was kindest to the hog.

They looked at the scales which would weigh each hog as it came through. Gates opened and closed automatically, either by computer control or electric eye. An inspector who was a plant employee observed the hogs going through both in natural light and by thermography. He would need only to press a button to take a picture of the animal along with recording its transponder number.

The company was very interested in whether any animal had bruises or injuries. The condition of each animal was recorded in the computer record along with its transponder number. The computer which would remember the origin and fate of each hog. Inside the plant, the employees would know if any hog had any abnormality which needed attention. Any problem which was found was traced back to the source in an effort to prevent any future repetition.

Even when the products left the plant lot numbers could be traced back to transponder numbers and finally to an individual pig. One of Inspector Winslow's duties was to review video tapes of the hogs at random to check on the performance of the plant employee inspectors. To a large degree, inspection was run by the plant. It had been privatized.

Soon they arrived back at the kill floor near the shackling pen. It was still called that although it was completely mechanized. No person needed to touch a hog.

There was something else Randal Sharp noticed. Years before there had been huge exhaust fans pulling air out of the plant. Into the plant came air from every opening. Back in the old days, if you opened a door to the outside, a hurricane seemed to try to drag you back into the plant along with dust and dirt from outside. Here all the air was pulled into the plant after first being filtered and cooled. If you opened a door air would rush out, not in. But most of the air stayed in the plant and was continuously cleaned, cooled, and dehumidified. Nowhere that he looked was there a wet or damp surface, or standing water.

Winslow and Yardley had returned from the government office with the program disk which had been cleared, "certified", they called it. They handed the disk to Mr. Long and he unlocked a door on the main computer and inserted the disk, meanwhile removing the other disk with the suspect program. He set the program in operation, allowing for the lost time as well as time for the few employees to get to their stations.

Mr. Long explained, "A few years ago when we had more business than we do now, we ran three seven hour shifts, with a hour to clean up. Now automatic machine washers with high pressure water and detergent do the job. Those washers are parked over there on a side rail where you can't see them now. They have the entire kill floor memorized and know what it should look like when it's clean. It isn't that they are so much faster but they do a much better job than a man...or woman.

"We inspect to see if the machine is doing the job. If it does not, if we find any dirt, then we find out why the machine has not done the job before we continue. The inspectors look over our shoulders, and do independent checks but it has been a long time since they found anything we had missed."

"That's right," concurred Inspector Winslow, "only one time in the past year that I remember. The detergent was colder than the formula called for and the weather was unusually cold. I think that particular problem has been solved."

"We keep it pretty cool in here anyway," remarked Mr. Long. "If you haven't been on a kill floor for thirty years, you will see a lot that's different. All those nooks and crannies where meat scraps could accumulate are gone. The washer knows when a surface is clean because the robot can see what it looks like and compare it with what it should look like. It has the whole plant memorized, so it knows where to point the soap gun."

"What it this floor made of?" asked Sharp "It sure doesn't feel like the old brick or concrete I remember."

"It isn't. It's called rubberroid and is a kind of plastic. Notice it isn't very slippery even when wet and is much softer to walk on...for man or hog. They just treat the surface they want to put it on and then just spray it on. It will stick to steel, concrete, wood, you name it. It even sticks to itself. It's not so easy to remove, though. It's very tough and the best way to remove it is to melt it."

Yardley asked if he could see some of the hogs going into the slaughter pen. He looked and looked but it seemed he did not find what he was looking for. When Colonel Sharp asked what he had been looking for he merely muttered that he was looking to see if the hogs seemed all right. To himself, Yardley thought that there might have been a paint mark on their necks but there was none. Not even anything that would rub off on his fingers when he rubbed their necks and heads. Might there be something that the machines might be able to see but which he could not since it might be an invisible color like infra red he mused?

They stopped and looked at the stunner. It was called the snake by the employees. It always gave Winslow the creeps to look at the stunner. It hung down from the ceiling, as limp as a green rubber hose. However, when it was energized it could move like a coiled serpent, only faster.

228

Inside were several electro-plastic muscles made of vinyl fluoride along with two electric cables leading to four sharp prongs that projected from its tip when it struck. Ordinarily the electric fangs were retracted in the head. Fiberoptic cables led to electronic eyes in the box that controlled it. It could recognize the neck of a hog and striking swiftly could touch a hog on the neck delivering a lethal shock. It was not supposed to touch anything else. If it struck the wall or some other object the electric prongs could be broken or the eyes damaged.

It reminded Winslow of a green mamba, the dreaded Dendroaspis viridis, the arboreal cobra, of the African jungle where some of his ancestors had once lived. Only it was larger and longer, and faster. Its strike could not be seen except as a blur.

Mike Long had no fear of it, but as a precaution would turn it off when working near it. The higher voltage to the electric fangs could be cut off while the mechanical functions would still work. It had not occurred to him that it could deliver a directed lethal blow even if its electric fangs were temporarily pulled.

Sometimes it was necessary for the inspector to observe it in action at close range to satisfy the requirements of humane slaughter directives.

Every so often Winslow would hear a bit of the opera. Just at that moment, three gangsters were singing a patter song, "Ese Cosa Nostra," in which each describes his particular form of criminal activity in low Sicilian dialect.

"Who is in charge of writing programs here?" asked Director Yardley.

"It depends upon what the program concerns," answered Mr. Long. "There are three of us involved. I usually write kill floor program changes. So does our engineer, and our financial officer. But sometimes we swap

229

around. None of us like to get into anything very complicated."

"Well, we need to get all three of you into the office and read out the kill floor disk. Then run a simulation, or real time if it doesn't appear too hazardous."

Not too hazardous meant that if they actually put the program into operation that no great damage to property could take place. It had not occurred to them that a program might be dangerous to life and limb.

"Getting us all here at the same time may be difficult, since the engineer is on the day shift, and the financial officer, Mr. Kahn, is away on vacation. He ought to be there in Mecca by now. Let's just run the program." Mr. Long started to say that he and Mr. Kahn did not get along well and often argued about plant matters but decided to let someone else tell them if they had to know.

"Is someone serving in his place?" asked Sharp.

"Only his secretary, who does all sorts of other reports. She is incidentally, my wife," remarked Mike Long.

"I'm sure she is able to attend to the ordinary run of payroll and purchase orders. The annual report requires all of us, it seems." Mike laughed. The plant was part of a much larger organization, but reports, and purchases, and payroll were handled locally with the information flowing upward in the organization. It was part of decentralization.

In a few minutes the first hogs were being brought out of the scalding vat and each animal's number from the transponder was printed onto a tag and placed on the carcass. Inspection no longer required this, but the plant had been set up this way years before. While the hogs were in the scalding vat, ultrasound probed and examined them looking for any defects. Any calcification, tumor, or abscess was noted and the hog was scheduled to be examined and trimmed. At the rail out station a plant employee trimmed out ordinary small abscesses and other blemishes. He could look at a video monitor that showed the location of the problem as seen by the ultrasound and the computer. The computer listened as the men spoke into their helmet microphones describing the lesion or blemish removed. More serious problems called for the inspector and the company veterinarian to do a post-mortem exam together. This was recorded on videotape and was later examined by a government veterinarian. Sometimes if there were serious problems or doubts about the health of a group of hogs (which was quite seldom), there would be a company veterinarian along with a company trimmer or the foreman, one or more government veterinarians and even a biologist or two from the farm, all crowded together at the final station, all talking, arguing, and examining hog viscera.

Winslow could only remember two such times in his seventeen years on the inspection force. One involved a new virus suspected of being transmissible to people, which turned out to be a genetic defect in a group of hogs and the other was an outbreak of Pasteurella in hogs that had been shipped too far. There had been a lawsuit between companies. He had been very glad to have been shoved aside and to "let the big dogs fight".

He had been watching the kill floor from an observation point, for there was little or no space for a person not actively working, and listening to the opera.

It was the scene in which Michael's Sicilian bride is killed in an assassination attempt on him. Michael and Apollonia sing a duet pledging their eternal love for each other. She breaks away from him and leaps into the car. As she does so Michael sees Fabrizzio fleeing from the explosion he knows will come. Michael cries, "No, no, aspetta!" but it is too late. There is a great roll of kettle drums and the Sicilian Idyll is over. His bride has been murdered. End of act two.

231

When we next encounter Micheal he has become a hardened Mafioso. But during the next act his brother is killed at the toll booth. This the famous ballet scene which gives the opera its special flavor. The godson, Don Carlo, has betrayed the brother to a rival gang.

The Opera was interrupted again briefly when the visitors had gone to the government office and attempted to run a comparison of the revised uncertified program with the older program. This had to be done via a modem linking up with another mainframe computer far away in Nashville. It was not easy but they isolated the differences and attempted to view a simulation. The result was inconclusive. One immediate problem was getting around the security system which allowed only certain authorized persons to program the company's computer. Naturally they wanted to know who had authored the changed disk. The signature did not match any of the known authorized programmers but that could have meant that the signature had been introduced earlier as a password.

Yardley and Sharp had come to some sort of conclusion. They had no grasp of why the program differed from the certified program. They had identified a mention of the color red. Perhaps it was intended to prevent the snake from becoming contaminated with blood.

Actually running the program in a limited way would give them the best answer. They stood before the shackling pen. Winslow and Long were behind them. There was a brief instant in which Winslow noticed that Sharp appeared to be the only white man in the group. It bothered him a little that the policeman Sharp outranked even Yardley. True Sharp was older than the rest of them by perhaps at least decade. Now the country had changed. The presence of police often struck real fear into the hearts of ordinary honest citizens.

He gave only a moment's thought that his great grandfather had been first a colored man, then a Negro, his grandfather a black, and his father had been called an African American. Then after terrible civil war in Southern Africa had sent thousands of refugees into the Americas, African Americans might be also be of Dutch, English, or East Indian, as well as Zulu, Basuto, Bechuana, Swazi, or other origin the term had faded. He was simply an American. DNA probes and other studies had made it possible for anyone to find out whom they were really related to in Europe, Africa or elsewhere. Some things were better, not worse. Americans were far less race oriented than they had been a century before.

Neither the police colonel nor the director seemed to be carrying a weapon, although appearances might be deceitful. It was not widely known, but most police did carry a transponder, not quite the same as those in the hogs, but a device that relayed information to the police vehicle and thence to a central office, revealing Sharp's location and condition, and recording what was said.

Total gun control had quite logically disarmed the police, just as it does in a prison. Prison guards do not carry weapons when they mingle with the prisoners. Police guns could not be stolen if they no longer existed. All of the good reasons for disarming the law-abiding public apply even more strongly to the police. Of course what is a weapon is in the mind rather than what is concealed in the pocket. That is, common objects become weapons when harm is intended. Sharp did carry a video camera to gather evidence and had used it around the plant. He placed the camera on a its tripod and turned it on. Then, having assured himself that it was recording he nodded to Dr. Yardley, who began to speak.

"Turn on the shocker, Mr. Long, but stand back well clear of it. Oh, and take off that red bandanna first and put it away in your pocket."

"The bandanna, why?" asked Long.

"You will see in a moment."

Long took off the red bandanna and then turned the stunner on. Part of the machinery had been turned off. It had taken some doing to get part of the plant to run alone with another part turned off. But it was part of their training to know how to do it.

A few moments passed. A warning light overhead that should have turned red indicating the system was on and dangerous stayed yellow and blinked indicating a malfunction and that the machine was off. Dr. Yardley took an apple out of his pocket, held it up so everyone could see it. It was nice and red. He tossed it out across the pen. Instantly, the snake struck it breaking it into tiny pulpy fragments. It reminded Winslow of a time in his youth when he had used a baseball bat while his brother pitched an apple in place of an expected baseball.

"My God!" exclaimed Long.

"Murder by proxy seldom thrives," remarked Yardley paraphrasing Shakespeare.

Mike Long was quite naturally stunned.

"That could have been me, or my red bandanna. It wouldn't have mattered whether the juice was on, but I bet it was. Lets see if it does it again. Here we can ball up the bandanna and try again."

"Go ahead, but I don't think it will strike this time."

Long tossed out the bandanna. The snake hung limp and innocently as if disconnected.

"If the stunner had failed to operate when it was supposed to, what would you have done?" Sharp asked Long.

"Why, I would have gone into the stunning area to try to fix the problem, and I would have found the yellow light on and ..." his voice trailed off into silence as he realized what could have happened.

"Let me explain," said Yardley. "The program needed to run only once. One time. Then the program erased itself and it would seem that you had some freak kind of accident. And if Inspector Winslow had not pulled the program and submitted it for examination just recently as a matter of luck," he hesitated, "you would not be here with us now.

"We first thought that there was some sort of red paint mark used on the hogs to present a target for the stunner. That's why we looked at the hogs and were puzzled. The red mark wouldn't have been a legitimate way to kill hogs, but we have discovered such goings on during times when equipment wasn't working quite in the prescribed manner. Those stunners are sturdy. Despite what the manuals say about taking care of them and not letting them hit hard objects they are not easily damaged. I'm sure that apple didn't hurt this one.

"One time we found hogs were being killed by a snake breaking their necks rather than correct electrical stunning. You were surprised at how hard it could hit. Proper reflexes programmed into the snake stop it when it touches the hog's neck.

"The next question is Colonel Sharp's to ask."

"Who would want to kill you and why?" asked Sharp. "You may remain silent if you wish. You may request council. But as of now we have no reason to hold you except as a witness.

"We have been informed that the financial officer has left the country, to go to Mecca on a pilgrimage. We will check to see if he has arrived there. If not, it will be suspicious.

235

"We will also need to find out other information which will bear on the case. We know there are only a very few persons who had access to the program. So far they are the engineer, your wife, and the financial officer. We will also alert the company to check into the financial records to see that all is well there."

"Now our problem is whether to pretend that nothing out of the ordinary has happened, or whether to pretend that there has been an unusual freakish fatal accident in the hope that such false information will help reveal the perpetrator. What do you think, Dr. Yardley?"

"Well you have or can soon get whatever authority to sequester (meaning house arrest) those who know anything about this or who might be involved. When we came here, we did not expect to encounter something quite like this. I will call in and get some help. In the meantime, we should probably let the plant operate as normally as it can even though it is now an hour and a half late. Would that meet with your approval? We do not want to get the public involved."

He meant unaware. Crime stories were kept out of the press and were only given out on a need to know basis. It reduced the incitement to violence and the imitation of crime. Stage, screen, and literature promoted the idea that people reasoned their way out of conflict. The law required this and it could be enforced.

Yardley and Sharp went back to the government office, and started making their calls. It was an odd hour to start doing such things in the wee witching hours between midnight and sunrise, but for many people the conventional diurnal cycle had disappeared. They lived in huge enclosed cities, for all the world as if underground, in artificial light. The world would soon come to a single time zone, GMT, or World Standard Time as it was now called. Plans for the change over were well underway.

Winslow stayed out of the office. There wasn't much room and the police would need the facilities contained there. There was nothing Winslow could do which Dr. Yardley would need help with.

The final act of Il Padrino was on. There was the baptism of Don Michael's son, with the killing both inside and outside the cathedral, harking back to an incident in Italian history in 1478 when the Pazzi and the Medici fought and killed each other in the Cathedral of Florence.

Don Micheal then seizes complete power with considerable skill. He identifies and has the traitorous Don Carlo and Tessio killed. Now he is supreme and the cycle is complete. He is now Il Padrino. He sings the same soliloquy that began the opera. The curtain closes as the orchestra shifts to the thunderous finale "O Fortuna" borrowed from the Carmina Burana.

"The curtain closes," thought Winslow. "I wonder if we will ever hear it performed publicly again." He went back to the office. Both men were still there. Others had joined them. These were men in black uniforms like the uniform worn by Colonel Sharp. One of them said to Winslow, "You will be free to go home when the shift is over. You are not to discuss any of this with anyone. You must wear this."

The officer held out a bracelet that had the appearance of a wristwatch, small, sleek, and efficient. It was a location monitor transponder, like those worn by Yardley and Sharp. It could tell where he was, and how he was, and it could listen in to any nearby sounds, even his pulse. If he were attacked or rendered unconscious, or made any unusual exertion it would be recognized.

"Leave it on until we ask for it back. It is totally waterproof and nearly shockproof. If you need help, press the little red dot. You can talk to us through it if you need to."

237

There was some comfort in the thought he was being treated as a fellow policeman rather than a suspect, which he might have been. Yardley and Sharp looked at him.

Yardly spoke, "You better get home and get some sleep, we will be getting back here later today. Right now we mean to get some sleep and let these other gentlemen take over. We don't know how big this thing may be."

It was an ominous reference to the possibility that the attempted murder involved neither greed, nor personal enmity, but a kind political terrorism that had begun a generation before and while it had been heavily suppressed, it still flared on occasion. Mike Long was in his position as kill floor foreman most directly responsible for the death of the pigs which were killed at the plant. It was this possibility that caused Randal Sharp to be called into the investigation.

It was still dark outside. It had been a short night's kill. He passed Mike Long on the way out. Mike was wearing a transponder too.

He remarked to Winslow, "You too?" Surprise showed in his voice.

Winslow almost remarked that the others, Yardley and Sharp had them. Instead he said, "You might still be in some danger. They want to know if you have trouble."

When he got home he was both very tired and yet not able to sleep. Someone had slipped a handbill into his newspaper. It upset his wife. Such things were clearly illegal. They violated the censorship regulations.

It read:

"Let's abolish animals! Animals compete with us for land, food, and resources.

Money, time, land, and effort are spent on lower animals which could and should be spent on human welfare and uplift.

Animals carry dangerous diseases and have disgusting habits. They present us with moral dilemmas and are the focus of much human neurotic behavior.

Worst of all, they remind us of our own humble and horrid origins.

Let's do away with the Park Reserves and open them up for human exploitation and settlement. We don't need animals."

It had bothered his small son, who tearfully asked, "Are they going to take Poopsie away?"

Poopsie was a small dog of general dachshund ancestry. Poopsie did bark a bit, but fortunately, the apartment was well soundproofed from its neighbors. It had been a bit hard to get a license to get Poopsie. Permissive pet licensing was no longer allowed. You had to have a reason for getting an animal. Companionship was not enough. After all, "Dogs are for biting people and are dangerous predatory carnivorous beasts that should not be kept around the house by the average citizen," the license bureau would say. Poopsie was a small concealable foreign dog who might just lurk under a table or chair and then leap out and bite someone. If something happened to Poopsie, he could not be replaced, at least not legally, so the boy's fears were not groundless.

The handbill bothered Erasmus Winslow too. Such handbills were ostensibly clandestine but were actually often trials of public opinion and were somehow done with the approval of the government. It wasn't not widely known but the population of North America was soon to reach a half billion. The U. S. alone was over three hundred ten million. Mexico's population was at one hundred twenty million. Canada had fifty million inhabitants.

During the previous generation he had seen the government begin to control the media. It had not been hard. It was rather like the way the conquistadors in Peru

239

and Mexico had taken over the very top of the Inca and Aztec empires. A tiny number of nobles had ruled the empires absolutely. A small number of persons of great wealth and power controlled the media. They had been deposed and or subverted. If you now asked the man in the street what pornography was he would have replied that it was violence or destructive criticism, especially of government or other large accepted institutions. It had earlier been referred to as political pornography. It hadn't taken long; no longer than it had taken to change the meaning of gay from happy and frivolous to homosexual.

He would have seen no conflict with the Bill of Rights; no more than a previous generation had with total gun control. This was what bothered him about the transponder. Though he had no doubt that they could have listened in on him at anytime, or accounted for his movements had they wanted to at any time in the past, now he was certain that they were listening and watching. He would not be able to scold the boy if the boy disobeyed, nor could he have sex with his wife. At least they would not be private, even though they generally were legal. He pointed it out to his wife, and then had to say that he couldn't tell her anything else about it or what had happened at the plant. They would be watching and listening the way he used to imagine God listening when he was a small boy, as his son now was.

Transponders and recording tape machines that recorded what machines did had been in aircraft long before his time. They eventually went into trucks and cars. A vehicle could be traced in its movements. It drastically reduced vehicle theft in a very short time. If a transponder were shut off the police immediately wanted to know why. Usually it was because of a very serious accident. The recorder tape would be impounded and the police and the insurance company would decide who was at fault. You could not argue that you had stopped for the stop sign if the recorder said you hadn't. Or if you had stopped and the other vehicle had not, you had proof. So it was a fair and neutral judge. More often it recorded all the vehicle functions and intermittent problems that were easily identified by the mechanics and electricians who serviced and repaired them, so they were accepted by the public.

In like manner, the transponder which snooped on Winslow's home life would bring the police running to defend him if he were attacked, or had a heart attack, or had to summon help quickly. It was in a sense an honor and a burden. Had he shown it to his son, the boy would have said, "Now you are a real cop!"

He himself thought in his heart of hearts that it was a disgusting sort of idea that would soon lead to terrible abuses. He had been brought up as a Baptist, and a Christian, and remembered the Biblical admonition that if we were not ruled by God we would be ruled by tyrants. It would be a terrible world when people were good only because they were closely watched and were forced to conform.

Who could watch the watchers and the enforcers?

Cops seldom did the sort of dramatic things that used to fill stage, screen, and literature. It was mostly dull work. It was viewed as preventive rather than punitive. Liberals had rehabilitated the term "Police State".

Lying in bed, Winslow wondered not just what the morning would bring. He wondered whether the public would decide that it was wrong to raise animals for food. Surely if it was wrong to hunt them, it was wrong to raise and kill them for food. If they did decide that it was wrong, how many would have to make such a decision? Would they be consistent and quit buying and eating meat?

He felt sure that food could soon come from the chemical industry rather than the factory farm.

241

In Asia they had eaten dogs and cats. He thought they still did, though on a very small scale. How did an Asian react to the idea that it was normal to kill and eat pigs but not dogs...or whales and dolphins? Would future generations decide that meat eating was somehow wrong and would they then think of him as some kind of monster like the slavers who had brought his ancestors to America? And if we should not hunt nor eat meat was it reasonable to preserve meat eating predators from extinction?

Such questions could not be asked or discussed in public. He could not explore such questions without getting into difficulties. If you questioned the existing order, things could happen to you...or your family. He pondered the power that the police, or someone over them had acquired. Who were they? Surely not the political figures he saw on the video. Had they planned his future in some general way, his unalterable fate? Who was allowed to present and promote ideas? Who wrote all those letters to the paper that always seemed to agree with the consensus, the government, and the editors? What had ever happened to all those dissenting small voices that used to be?

Winslow was an educated, literate man. He had read history when people still read books. He had learned history. It seemed natural to one who liked literature and music, especially opera. The old books were crumbling because of the acid paper they had been published on. More often if he wanted to read something from the library, it was not in the form of a conventional book but was delivered to the memory of his personal computer via fiberoptic cable which also connected to his telephone and video system. Then he could read it at leisure on a small screen he could hold on his lap.

Now it seemed that titles that he had once read were always away being recorded on disk, and were unavailable, or were lost, or had been classified as not available to the general public because of dangerous content.

He got up. He had heard his son stir. The boy wanted a glass of water. He got a glass of cold milk for himself and the boy. They both sat on the edge of the bed and looked out at the cold pale moon.

"Daddy, do you remember when the men landed on the moon?"

"No, Teddy, it was long before I was born."

"Will anyone ever go back? Why would they go so far away?"

He started to say, "I hope so," but corrected himself and said, "I'm sure they will. Men will go much farther than that looking for freedom." He did not care that others would hear him.

He added, "And our animals, our pets like Poopsie, will go with us to those other worlds. They have paid the price of their tickets many times over."

There was a sort of epilogue. Before the week was over, the police had again interviewed Mr. Long and Inspector Winslow. They retrieved the transponders. Lieutenant Sharp called them into the office and told them that although Mrs. Long had been suspected, she had been cleared. Evidently they had known that Mr. Kahn and Mr. Long had both been suitors for her hand at some time in the past. Perhaps jealousy had played a part in the attempted murder. Perhaps ideology for Mr. Kahn had been found to have some connections with one of the animal liberation terrorist groups. In addition, funds had been absconded via the computer and some of the company books had been altered to hide this diversion of funds. Most suspicious was the failure to return. From Mecca he had gone into a new African country which had no extradition treaties. There was no political terrorist plot so far the police could tell...or were willing to tell.

It seemed a banal anticlimax and they both said so. Sharp totally agreed. Evil is very often banal. They were not to discuss the matter. It would remain closed. Sharp had told them more than he was supposed to but it seemed reasonable to let them know what dangers might remain if any.

They looked at each other, each wondering what the other was thinking. But none of the three of them were willing to risk revealing their private thoughts so they simply said goodby to Colonel Sharp and went about their work.

The Cast of Characters

A cast of characters is listed in order of their appearance on stage, with a few notes about them. All are fictitious unless otherwise noted here. Some of the incidents are true. Some are not but make good stories. In such cases the names have been changed to protect the guilty and the innocent.

Dr. George Patrick Sullivan was born in 1887, and spent his early years in an orphanage near Boston. He died in 1963.

Dr. Wilber Smith was born in 1911, died 1996. He was a graduate of TEXAS AMC School of Veterinary Medicine, class of 1933.

Inspector Boswell was a lay inspector at K. C. Packing who first helped Smith learn inspection.

"Lampwick" Garrigan was the cattle and hog buyer at Old Nasty. Garrigan had come from a good family back east, but developed bad ways during the Great War and had been disowned by his family...not that he gave a hoot.

Dave Lightfoot was the cattle and sometimes hog driver along with Lampwick. He was not too bright a fellow. Lampwick, who was his supposed friend, often got Lightfoot into trouble just for fun.

Bill was the kill floor foreman at Old Nasty.

Inspector Reilley, was a lay inspector at Old Nasty, in the same circuit in Kansas City with Smith, Sullivan, and Boswell.

Lazar Weissmann, was an immigrant from Jassy, Romania, who handled the business office at Old Nasty, of which he was a major owner. The business also included loan sharking and other unnamed criminal activities.

Attila Farcas, was the other owner of Old Nasty, an immigrant from Transylvania, stick man at Old Nasty, and sometimes foreman as well as part owner.

Ezra Hershfeld was Elsie's father, an Illinois dairy farmer. He was also known as Grandfather Hershfeld, Timothy's grandfather.

Susie was a pet sow belonging to the Hershfeld family.

Dr. Anna Paluski was a Polish born veterinarian, who in turn was married to; Wadislaw Kozincky, a pilot in the Polish Air Force in 1939, Gerhard Stein, a German officer who rebelled against Hitler,   and Colonel Stanley Wallace, an American soldier who rescued Anna from the Ravensbruck concentration camp.

Dr. Heinrich Kruger, a German veterinarian, was born in Hamburg, served in the Wehrmacht in the USSR during WWII, worked in the US in Chicago and later in Wisconsin, where he died about 1976.

Felix was the editor of an unnamed newspaper in the Chicago area.

Olympia Westchester was the pen name of Hanna Baruk. However she had several husbands and consequently other names. She was a very successful newspaper columnist and fashion reporter.

Kowalski Brothers was the name of a meat packing plant near Chicago owned by the Kowalski family. Micheal worked the kill and was generally involved with inspection. Steven worked the loading dock and the business office.

Angel and Soul were cats who lived in Kowalski plant, whose job was to catch rats and mice.

Dr. Mullins was the veterinarian assigned to review Kowalski Brothers.

Dr. Demitrious Christophoros was the veterinarian in charge at Northport Packing.

Inspector Jones was a lay inspector there who was on antemortem inspection.

Rabbi Jeheudi Bekelstein was the schlacter at the plant and was responsible for kosher killing of sheep and cattle.

Dr. Ben Levinsohn, was born in Israel, received his veterinary training in Ontario, and was in the process of immigrating to the United States, via Canada.

Sam Mordacai was general manager of Northport packing.

Sparky was a young Yorkshire boar who turned up missing after a load of pigs was sent to Weiss Packing.

Ben Sorensen was a hog farmer from lower Wisconsin. His son was Joe. They were searching for Sparky.

Mr. Johnson was the weigh master at Weiss Packing Company.

Ed Weiss was co-owner of Weiss Packing Co.

Dr. C. J. Fuller was a Washington official in charge of chemical additives.

Carol Foreman was a very real person about that time.

Inspectors Jose Ricaro and D. E. Williams, were inspectors at Pintafour Packing, with Dr. Smith.

Mr. Halley was the plant manager at Pintafour Packing.

Dr. Karl B. Delancy was Circuit Supervisor and later Regional Supervisor.

Randal Sharp was a Colonel in the National Federal Police, the son of Preston Sharp, nephew of Horatio and Beauregard Sharp, the owners of Sharp Brothers' Packing, and grandson of H. H. Sharp who founded the business. Having seen that there was no future for him in the family business, he went to law school and specialized in criminal law. He then served in the military. Finally when all police were federalized he was moved into the Federal police.

D1270483